Wicked Problems

Wicked Problems

The Ethics of Action for Peace, Rights, and Justice

Edited by

AUSTIN CHOI-FITZPATRICK,
DOUGLAS IRVIN-ERICKSON,
AND
ERNESTO VERDEJA

OXFORD
UNIVERSITY PRESS

OXFORD
UNIVERSITY PRESS

Oxford University Press is a department of the University of Oxford. It furthers
the University's objective of excellence in research, scholarship, and education
by publishing worldwide. Oxford is a registered trade mark of Oxford University
Press in the UK and certain other countries.

Published in the United States of America by Oxford University Press
198 Madison Avenue, New York, NY 10016, United States of America.

Library of Congress Control Number: 2022930113

ISBN 978-0-19-763282-6 (pbk.)
ISBN 978-0-19-763281-9 (hbk.)

DOI: 10.1093/oso/9780197632819.001.0001

3 5 7 9 8 6 4 2

Paperback printed in Canada by Marquis
Hardback printed by Bridgeport National Bindery, Inc., United States of America

Chapter 1 is excerpted from Tony Gaskew, *Stop Trying to Fix Policing: Lessons Learned from
the Front Lines of Black Liberation* (Rowman & Littlefield, 2021), and published here by
permission of Rowman & Littlefield. All Rights Reserved.

To our teachers and our students,
and to those who never lose faith
in the struggle for peace and justice.

Contents

Acknowledgments

We are enormously grateful to the contributors to this volume, who helped us take a small idea and expand it into a rich, nuanced, and complex set of reflections on some of the key problems of our times. We are also grateful to our colleagues at our respective institutions: University of San Diego's Kroc School of Peace Studies, George Mason University's Jimmy and Rosalynn Carter School for Peace and Conflict Resolution, and University of Notre Dame's Kroc Institute for International Peace Studies and Department of Political Science. This project is much better for having emerged from decades of conversations with colleagues, students, collaborators, and antagonists, whether in hallways and classrooms, on the ground, or in the streets.

We are grateful for the technical and logistical support we received at each of our institutions. At the University of San Diego, Marjon Saulon served as our tireless copyeditor and Angelica Sanabria served as our enthusiastic marketing associate. The bibliography was compiled by Christina Grossi at Notre Dame, and Emily Sample and Ziad Al Achkar provided feedback and advice on the manuscript at George Mason University.

At Oxford University Press we are grateful to have worked closely with Angela Chnapko. From our first conversations about this project, Angela championed a book that spoke to puzzles that are both contemporary and enduring. We would also like to thank the excellent production team at the Press, including the copyeditor Judith Hoover and Vinothini Thiruvannamalai of Newgen.

Austin Choi-Fitzpatrick would like to thank for their support colleagues at the University of San Diego's Kroc School for Peace Studies, the University of Nottingham's School of Sociology and Social Policy, Harvard University's Kennedy School of Government, and University of California—San Diego's Institute for Practical Ethics. In particular I would like to thank the folks building, resourcing, and defending ecosystems where free-range scholarly inquiry can take place: at USD, Dean Patricia Marquez, Provost Gail Baker, and President Jim Harris; at Nottingham, Rights Lab Director Zoe Trodd, Research Director Kevin Bales, and

Pro-Vice-Chancellor Todd Landman; at Harvard, Carr Center for Human Rights Director Mathias Risse and Executive Director Sushma Raman; and at UCSD Associate Dean John Evans. Without their defense of the liberal arts, academics like me (and perhaps books like this one) would be thin on the ground. During this project I benefited from conversations about dilemmas with Kroc, Rights Lab, and Carr Center colleagues, as well as Gordon Hoople, Ev Meade, Amelia Watkins-Smith, Maritza Johnson, Sanaz Mobassari, Maximillian Kasey, Emily Keehn, Minh Dang, Kevin Bales, Joshua McIvor-Andersen, Daniel Leonard, Zoya Sardashti, Jordan Siegel, Jason Ahn, and Jesse Cusic. Emily Keehn, Ginger Hallack, and Louise Lue provided incredible administrative support over the course of this project, and this book is much better for the obstacles they cleared from my path. Perhaps this book should have been written by Jenny Choi-Fitzpatrick, who is a professional dilemmatician and determined finder of common purpose and higher ground. I'm grateful for everything I learn from her. Our vectors—Eden Justice and Aila Pax—are perhaps at the very moment thinking of new identities, incentives, and ecosystems for a more just and peaceful world. All this is really for them.

Douglas Irvin-Erickson would like to thank Yasemin and Troian for their generous and joyous spirits, and for making a book about peace and a more just future seem like a worthwhile project. My colleagues at the Carter School were especially helpful. Thank you to Susan Allen, Kevin Avruch, Charles Chavis, Sara Cobb, Suzanne de Janasz, Leslie Dwyer, Thomas Flores, Marc Gopin, Susan Hirsch, Karina Korostelina, Tehama Lopez Bunyasi, Terrence Lyons, Patricia Maulden, Alpaslan Özerdem, Agnieszka Paczyńska, Arthur Romano, Daniel Rothbart, Richard Rubenstein, Mara Schoeny, Juliette Shedd, and Solon Simmons.

Ernesto Verdeja would like to thank colleagues and friends at the Kroc Institute, who have long been intellectual partners in tackling the challenging and seemingly overwhelming problems our societies face today. I would especially like to thank Asher Kaufman at the Kroc for sustaining and expanding our rich intellectual environment, as well as colleagues Ashley Bohrer, Catherine Bolten, David Cortright, Robert C. Johansen, John Paul Lederach, George A. Lopez, Ann Mische, Laurie Nathan, Atalia Omer, Jason Springs, and Peter Wallensteen, who have all helped me think through the issues in this book in greater depth. Outside of Notre Dame, conversations with Joyce Apsel, Bridget Conley, Christian Davenport, Larissa Fast, James Finkel, Tibi Galis, Jocelyn Getgen Kestenbaum, Ashley

Greene, Alexander Hinton, Kristina Hook, Adam Lupel, Deborah Mayersen, Hollie Nyseth Brehm, Renata Segura, James Waller, Edward Weisband, Kerry Whigham, Lawrence Woocher, and Andrew Woolford have all sharpened my own thinking on deep ethical dilemmas. And very special thanks to Bettina Spencer, who has always encouraged me to be an engaged scholar, committed to ethical change.

Contributors

Editors

Austin Choi-Fitzpatrick is University Professor at the University of San Diego's Joan B. Kroc School of Peace Studies. Most of Austin's work occurs in the fecund fissures between established fields, currently including social movements, human rights, science and technology studies, engineering, and public art. His books include *What Slaveholders Think* (2017) and *The Good Drone* (2020).

Douglas Irvin-Erickson is an assistant professor at the Jimmy and Rosalynn Carter School for Peace and Conflict Resolution at George Mason University, where he directs the Raphaël Lemkin Genocide Prevention Program at the Center for Peacemaking Practice. His books include *Raphaël Lemkin and the Concept of Genocide* (2017) and *Building Peace in America* (2020).

Ernesto Verdeja is an associate professor of political science and peace studies at the Kroc Institute for International Peace Studies, University of Notre Dame, and executive director of the nongovernmental organization Institute for the Study of Genocide. He works on the causes and prevention of genocide and mass atrocities, and consults with governments and human rights organizations on atrocity prevention and political reconciliation.

Authors

Donald Anthonyson is the director of Families for Freedom, where he was previously a lead organizer and board member and has led the efforts of the International Deportee Justice Campaign. Donald migrated to the United States in 1979 from Antigua and has been involved in various social issues, including police brutality (Elenanor Bumphus Justice Committee) and antiracial responses (NYASA) to immigration. Anthonyson's contribution appears in Chapter 7.

Jessica Baumgardner-Zuzik is the senior director for learning and evaluation at the Alliance for Peacebuilding. She has expertise in peacebuilding and monitoring, evaluation, and learning (MEL), with more than twelve years of experience in leadership positions in multilateral institutions and NGOs developing and implementing monitoring and evaluation approaches in conflict-affected and fragile states. She oversees AfP's learning and education portfolio, where she conducts research and works to improve capacity and understanding of MEL techniques.

Ashley J. Bohrer is a scholar-activist based in Chicago. She holds a PhD in philosophy and is currently assistant professor of gender and peace studies at the Kroc Institute for International Peace Studies at the University of Notre Dame. She is the author of *Marxism and Intersectionality: Race, Gender, Class, and Sexuality under Contemporary Capitalism* (2020).

Travis Akil Brookes works with Rethink Masculinity, a D.C.-based collaborative effort to address gendered violence by engaging masculine-identifying people in work to promote healthy masculinities. Brookes' contribution appears in Chapter 7.

Minh Dang is executive director and cofounder of Survivor Alliance, an international nonprofit focused on leadership development for survivors of slavery and human trafficking (www.survivoralliance.org). She is also a research fellow at the University of Nottingham's Rights Lab, studying mental health and well-being of survivors of slavery.

Philip Gamaghelyan is an assistant professor at the Joan B. Kroc School of Peace Studies at the University of San Diego. He is also a conflict resolution scholar-practitioner, cofounder and board member of the Imagine Center for Conflict Transformation, and the managing editor of the *Caucasus Edition: Journal of Conflict Transformation*. He works in the post-Soviet states, as well as Turkey, Syria, and other conflict regions, engaging policymakers, journalists, educators, social scientists, and other discourse-creating professionals. His research is focused on politics of memory in conflict contexts as well as on critical reevaluation and design of conflict resolution interventions.

Tony Gaskew is a professor of criminal justice and a faculty affiliate in Africana studies at the University of Pittsburgh, Bradford. As a critical race theorist, his research is immersed in dismantling criminal justice systems of oppression. His books include *Rethinking Prison Reentry: Transforming Humiliation into Humility* (2014) and *Stop Trying to Fix Policing: Lessons Learned from the Front Lines of Black Liberation* (2021).

Beatrix Geaghan-Breiner is an undergraduate at Columbia University studying the history of twentieth-century U.S. foreign policy. She is a Laidlaw Research Fellow interested in international politics and peace and has published on U.S. sanctions on Iran in the *Columbia Undergrad Law Review*.

alicia sanchez gill is a queer, afrolatinx survivor who has called D.C. home for almost twenty years. Her work centers on creating survivor safety without prisons. She has fifteen years of experience in cross-movement organizing firmly grounded in Black, queer feminist theory and lived experience. She is the executive director of Emergent

Fund, a national, rapid-response, social justice fund created with the explicit goal of supporting Black, Indigenous, and people of color organizers responding to the biggest crises of our time.

Felicity Gray is a PhD scholar at the Australian National University and a practitioner in protection of civilians and conflict response. Her research examines unarmed and nonviolent strategies for the protection of civilians in armed conflict. Her research spans Lebanon, Berlin, New York City, Myanmar, and South Sudan. She has worked as a practitioner of nonviolent protection action in South Sudan.

Susan F. Hirsch is the Vernon and Minnie Lynch Chair and Professor at the Jimmy and Rosalynn Carter School for Peace and Conflict Resolution at George Mason University. She is the author of *In the Moment of Greatest Calamity: Terrorism, Grief, and a Victim's Quest for Justice* (2009) and *Mountaintop Mining in Appalachia: Understanding Stakeholders and Change in Environmental Conflict* (2014) and, with Agnieszka Paczyńska, the editor of *Conflict Zone, Comfort Zone: Ethics, Pedagogy, and Effecting Change in Field-Based Courses* (2019).

Elizabeth Hume is the Executive Director of the Alliance for Peacebuilding. She is a conflict prevention and peacebuilding expert with more than twenty years of experience in senior leadership positions in bilateral and multilateral institutions and NGOs developing and implementing conflict prevention and peacebuilding programs in conflict-affected and fragile states. She oversees AfP's Policy and Advocacy program, where she co-leads the Global Fragility Act Coalition and leads other policy legislation reforms in the peacebuilding and conflict prevention sector.

Deena R. Hurwitz is a Schell Center Senior Human Rights Fellow at Yale Law School and a human rights consultant. She taught international human rights law clinics and seminars for close to two decades and is the founding director of the International Human Rights Law Clinic at the University of Virginia School of Law, where she was also a professor of law from 2003 until 2015. She has worked in and written on the Middle East and North Africa for many years. Her publications include *International Human Rights Advocacy Law Stories* (coedited with Margaret Satterthwaite and Douglas Ford, 2009).

Latishia James-Portis is a movement chaplain and trauma-informed facilitator. She works at the intersection of faith, sexuality, and reproductive and healing justice. James-Portis' contribution appears in Chapter 7.

George A. Lopez is the Hesburgh Professor of Peace Studies, Emeritus, at the Kroc Institute for International Peace Studies, University of Notre Dame. He has written fifty articles and book chapters and six books on sanctions. From October 2010 through July 2011, he served on the United Nations Panel of Experts for monitoring and implementing UN sanctions on North Korea.

Tim Murithi is head of program at the Institute for Justice and Reconciliation based in Cape Town and extraordinary professor of African studies at the Centre for African Studies, University of the Free State, South Africa. He has over twenty-three years of experience in the fields of peace, security, international justice, governance, and development.

Daniel J. Myers is president of Misericordia University. He previously served as provost at American University in Washington, D.C., and Marquette University and as professor of sociology at the University of Notre Dame.

Laurie Nathan is professor of the practice of mediation and director of the Mediation Program at the Kroc Institute for International Peace Studies, University of Notre Dame. His most recent journal articles have appeared in *Global Governance, Global Policy, Swiss Political Science Review,* and *Third World Quarterly.*

Reina C. Neufeldt is associate professor of peace and conflict studies at Conrad Grebel University College at the University of Waterloo. She works on applied peacebuilding ethics, reflective practice, and program monitoring and learning with NGOs. Her publications include *Ethics for Peacebuilders: A Practical Guide* (2016).

Nicole Newman is an organizer in Washington, D.C. She earned her MS in organizational development at American University and is the cofounder of Two Brown Girls Consulting Cooperative. Newman's contribution appears in Chapter 7.

Agnieszka Paczyńska is associate professor at the Jimmy and Rosalynn Carter School for Peace and Conflict Resolution at George Mason University. She is the editor, with Susan F. Hirsch, of *Conflict Zone, Comfort Zone: Ethics, Pedagogy, and Effecting Change in Field-Based Courses* (2019) and the author of *The New Politics of Aid: Emerging Donors and Conflict-Affected States* (2019).

Darakshan Raja is the codirector of the Justice for Muslims Collective, a community-based organization that works to dismantle structural Islamophobia through community organizing and empowerment, raising political consciousness, shifting narratives, and building strategic alliances across movements. Raja's contribution appears in Chapter 7.

Janis Rosheuvel is a program director at Solidaire, a major donor network moving money to progressive social movements. She also does antiracism training with the Flatbush Tenant Coalition. She earned her MA in conflict resolution from the University of Bradford, lectured at John Jay College, and served as a Fulbright fellow in South Africa. Rosheuvel's contribution appears in Chapter 7.

Kirssa Cline Ryckman is an associate professor at the School of Government and Public Policy at the University of Arizona. Her research focuses on the outcomes of both nonviolent collective campaigns and violent movements, as well as the use of violence against civilians in the form of terrorism and within the context of civil war.

Noam Sandweiss-Back is the director of partnerships with the Poor People's Campaign: A National Call for Moral Revival and the program manager with the Kairos Center for Religions, Rights, and Social Justice at Union Theological Seminary, New York.

Zuri C. Tau is a founder of LiberatoryResearch.com and a principal at Social Insights Research, LLC. She is currently earning her PhD in sociology from Georgia State University. She writes about and studies oppositional consciousness, women of color leadership, and social movements and is the managing editor of the academic journal *City and Community*. Tau's contribution appears in Chapter 7.

Liz Theoharis is the co-chair of the Poor People's Campaign: A National Call for Moral Revival and the director of the Kairos Center for Religions, Rights, and Social Justice at Union Theological Seminary. She is the author of *Always with Us? What Jesus Really Said about the Poor* (2017) and co-author of *Revive Us Again: Vision and Action in Moral Organizing* (2018).

Introduction

Wicked Problems—The Ethics of Action for Peace, Rights, and Justice

Austin Choi-Fitzpatrick, Douglas Irvin-Erickson, and Ernesto Verdeja

What kinds of ethical challenges emerge in pursuit of peace, human rights, and social justice? A desire to bring change should lead to action. Action quickly leads to complexity. And some of the most complex complexity involves ethics. Textbook ethics are a good start for how we think about the world, but have a "last mile problem." They help us get started, but when the moment of decision is reached, and when the rubber meets the road, as it were, judgment and discernment (or id, impulse, bias, and inertia) may have the final word. That's why we created this book—to hand to our students, our colleagues, and our friends, certainly—but to educate ourselves as well. We learned a lot from the chapters in this book.

For those of us who believe our goal should be not just to understand the world but to change it, this book proposes a simple question: *What kind of ethical challenges are involved in those change efforts?* The three of us teach in graduate programs where we invite our students to wrestle with this question and its many answers. But we know that the dilemmas of change-oriented action are most visible not in a classroom but in practice, as we go about the real work of trying to understand and change the world.

***

Let's start with an example. Nonviolence is a wicked problem. Not since the late 1960s have America's contradictions been on fuller display. A pandemic laid the country low, a national reckoning on race is afoot, and the stock market is booming. If America's streets are alight, so is its conscience. It seems everyone loves Martin Luther King Jr., which is a sure sign that he is poorly understood and that the radicalness of his message risks being lost.

Austin Choi-Fitzpatrick, Douglas Irvin-Erickson, and Ernesto Verdeja, *Introduction* In: *Wicked Problems*.
Edited by: Austin Choi-Fitzpatrick, Douglas Irvin-Erickson, and Ernesto Verdeja, Oxford University Press.
© Oxford University Press 2022. DOI: 10.1093/oso/9780197632819.003.0001

In fact, every organizer in Dr. King's wake has understood his commitment to creating conflicts that laid bare the hypocrisy of American peace. They know that America's prophet of nonviolence is also the standard-bearer for disruptive civil disobedience and mass noncompliance. Much has been made of King's principled commitment to nonviolent action. Rightly so. There is much more to be said, however, about the vital importance, indeed the vitality, of *conflict*. Too often nonviolence has been misread, by the public but also by those who should know better, as a kind of passivity that avoids disrupting the status quo and confronting illegitimate power.

Struggles for peace, rights, and justice have long been framed as occurring between two ethical poles. The first is represented by King, writing in *Beyond Vietnam*.

> "Peace and civil rights don't mix," they say. "Aren't you hurting the cause of your people?" they ask. And when I hear them, though I often understand the source of their concern, I am nevertheless greatly saddened, for such questions mean that the inquirers have not really known me, my commitment, or my calling. Indeed, their questions suggest that they do not know the world in which they live.[1]

The second pole in this ethical debate is pointedly formulated by a young Malcolm X, writing in "The Ballot or the Bullet":

> You may wonder why all of the atrocities that have been committed in Africa and in Hungary and in Asia, and in Latin America are brought before the UN, and the Negro problem is never brought before the UN. This is part of the conspiracy. This old, tricky blue eyed liberal who is supposed to be your and my friend, supposed to be in our corner, supposed to be subsidizing our struggle, and supposed to be acting in the capacity of an adviser, never tells you anything about human rights. . . . When you expand the civil-rights struggle to the level of human rights, you can then take the case of the black man in this country before the nations in the UN. You can take it before the General Assembly. You can take Uncle Sam before a world court. Uncle Sam's hands are dripping with blood, dripping with the blood of the black man in this country. He's the earth's number-one hypocrite. . . . Take it into the United Nations, where our African brothers can throw their weight on our side, where our Asian brothers can throw their weight on our side, where our Latin-American brothers can throw their weight on our

side, and where 800 million Chinamen are sitting there waiting to throw their weight on our side. Let the world know how bloody his hands are. Let the world know the hypocrisy that's practiced over here. Let it be the ballot or the bullet. Let him know that it must be the ballot or the bullet.[2]

This tension—between these poles and within this complex relationship—deserves more attention and is one of the many challenges to ethical action that led us to compile a volume entitled *Wicked Problems*. The two perspectives of King and Malcolm X we selected strike to the heart of this reckoning with conflict. They go to the very core of the deep ethical and practical challenges any change-oriented social movement must confront. While King and Malcolm X are often reduced to polar opposites—their ideas oversimplified into a neat binary where Malcolm X supports violence and King defends nonviolence—both encouraged their followers to think about violence and nonviolence in strategic and ethical terms, inviting critical reflection about when violence could be necessary and justified to secure the liberation of the oppressed. What's more, both understood the strategic role conflict—in both its violent and nonviolent forms—plays in liberation. Both understood that conflict as disruption is necessary for peace. Both realized that the struggle against injustice requires a complex series of judgments and actions that involve combining one's commitment to fundamental values with practical concerns over strategic and tactical efficacy, often in rapidly changing contexts where the outcome is unclear.

In "The Ballot or the Bullet," we find Malcolm X at a point in his life when he is thinking seriously about what kinds of institutional safeguards need to be in place for the Black nationalism movement in the United States to consider accepting American democracy as viable. Malcolm X had long advocated for Black economic self-sufficiency, pride for African Americans, and Black separatism within the U.S. political community as a response to white racial prejudice and the gap between American democratic ideals and the reality of American culture, society, politics, and economics. This Black nationalist project, for Malcolm X, always required self-defensive violence against white supremacy. In "The Ballot or the Bullet," he links this Black struggle to a global human rights movement where Black, Brown, and formerly colonized people around the world stand in solidarity with each other, willing to mobilize their global might in defense of their collective human rights. The United Nations, by this light, offered one path forward, as a forum for highlighting American perfidy and hypocrisy. The clear implication is that the UN should consider the United States a site for human rights intervention. The contemporary version of this argument is to suggest that an intervention based on the UN Responsibility to Protect is required to defend traditionally marginalized communities from human rights violations within the United States, since social, political, and economic systems have proven insufficient or complicit.

Both King and Malcolm X clearly saw nonviolence and the ballot as desirable means and ends, respectively. Both saw conflict as a potential resource for destabilizing the convenient truths of the white majority, even if they differed on the uses of violence. Rereading King and Malcolm X helps us to see that calls for nonviolence suggest tactics that increase conflict. Shedding fresh light on this history also reminds us that King and Malcolm X saw the struggle for the abolition of poverty, for Black liberation, and for civil and human rights in the United States within a global context. If we want to talk about "ideas in action," we must find and build links with local grassroots efforts as well as global justice movements and movements for decolonization.[3]

What is more, this framing of King and Malcolm X forces the recognition that conflict avoidance means siding with the status quo, siding with systems of injustice, and giving up on peace and justice. The same reckoning is happening worldwide, as societies everywhere face similar challenges over how to confront violence, injustice, and disenfranchisement.

Our argument is clear: Struggles for peace require conflict.

* * *

Struggles for peace create challenges. To explore this book's theme, what we're calling *wicked problems*, we have invited contributions from activists, educators, scholars, and scholar-practitioners. We asked them a simple question: *What kind of ethical challenges emerge in the course of doing the work that you care the most about?* You are holding their answers in your hand.

This book does not showcase long scholarly works on the nature of ethics as debated in the academy. This book does not canvas a few high-profile case studies on one narrow puzzle or dilemma. Rather, we have sought out reflections on the broad range of ethical quandaries faced in everyday struggles for peace and justice.

Often, the field of peace and conflict studies, as it is normally called today, in fact means something closer to peace and "violence" studies. Nevertheless, the idea of conflict (as disruption) functioning as a generative force for desirable change is central to some strands of peace practice and theory, a point we address in this introduction. Furthermore, the peace and conflict studies literature has considered the ethics of peacebuilding, transitional justice, and traditional conflict resolution. In particular, scholars have focused considerable attention on the ethical dilemmas of specific humanitarian interventions, whether they were justified and whether they achieved their intended outcomes. But interventions—just like negotiations, social movements, and peacebuilding efforts—are undertaken by *people*. Our attention is thus turned to the enduring dilemmas that peace and conflict resolution practitioners face in their everyday work. That is why we set out to hear more voices, about more dilemmas, in more contexts, and across more levels of intervention than is possible in more conventional scholarly books. It is also why this book embraces a heterogeneous range of ethical frameworks, empirical analyses, and perspectives. And that is why many of these short pieces are written in the first person, about a real dilemma faced by a real person, and in real cases where people are trying to show up and make a difference. This book might hail from a field called "peace and conflict studies," but we believe these stories resonate more broadly and may be relevant to anyone committed to aligning real-time change-oriented action with broader ethical principles.

Writing a book about ethical dilemmas in any field raises several questions, not the least of which are the following: *What is the field? What do we mean by ethics? And what do we mean by a dilemma?* There is a wide range of perspectives about the scope and content of ethical discourse across different domains of this field, especially related to practical and applied ethics, but our intention is not to adopt any one particular conceptualization. Rather, our objective is to depict ethics as an everyday practice that engages practitioners *as a matter of course*. We focus on ethics and dilemmas as practical puzzles experienced by practitioners in the midst of their lives rather than rarified units of philosophical abstraction puzzled over by our peers in academe.

Organizing Concepts: Field, Ethics, and Dilemmas in Peace and Conflict Studies

Let us begin with the question of how to conceptualize peace and conflict studies. We refer to it as a field rather than a traditional discipline because of its highly heterogeneous nature. The field nevertheless has, we believe, a nexus of shared understandings.[4] By "peace and conflict studies," we refer to several features of this nexus.

First, we are referring to the broad range of scholarly inquiry concerned with investigating the causes, patterns, and meaning of violence and conflict, as well as the spectrum of responses to mitigate or terminate violence and to prevent its recurrence. This requires examining the many understandings and justifications of peace—some at odds with one another—as well as available approaches and strategies to achieving it. All three of us believe contentious politics and disruptive protests are often required in pursuit of a more just and peaceful society.

Second, this field is interdisciplinary and multidisciplinary. It draws on methods and research from across the social sciences and humanities, and increasingly other disciplines as well. Within this broad definition, there are numerous methodological perspectives and assumptions that underpin research and practice.

Third, the field is explicitly normative in orientation, in particular with its focus on the reduction or elimination of avoidable human suffering and harm and the promotion and sustainment of peace. To generate knowledge in this field is to take sides.

Fourth, and important for our purposes, the field has a strong connection to practice. By "practice" we mean concrete action to reduce or eliminate harm and to enhance human flourishing. These efforts involve an enormous number of formal and informal actors, from the grassroots to the global. That's why many of the contributors to this volume are practitioners.

Fifth, the field has a complex genealogy. Peace and conflict studies draws on historical lineages that intersect in some places while diverging in others. Here we cannot provide a comprehensive history of the field—others have done so already[5]—but we can emphasize the oft-overlooked: this field has been deeply informed by popular social movements, whether they be against war and state repression, colonialism and anticolonialism, inequality and poverty, or discrimination and cultural destruction.

Finally, academic programs have played a crucial role as incubators of peace research and have developed a large body of scholarly knowledge. There is now also a large "professional" class of peace practitioners based in international organizations like the UN and the European Union, civil society nongovernmental organizations (NGOs) and social movements, governments, think tanks and academic programs. The significant experience of this professional class has resulted in a sophisticated corpus of theoretical, technical and applied knowledge in areas like conflict prevention, peacebuilding, and development. However, this technical and increasingly institutional expertise must also be tempered with the rich and varied practical knowledge and strategies developed by peace activists, honed through years of engagement in the streets, in community centers, in the courts, and in the halls of political and economic power.

Peace and conflict studies, then, consists of research, teaching, and practice in a constellation of related areas. Today these include alternative dispute resolution, antiracism work, civil rights advocacy, civil society building, collective action, contentious politics, counterterrorism, dispute resolution, genocide and atrocity prevention, human rights, human security, humanitarian intervention and response, international law, mediation, negotiation, nonviolent direct action, peace education, peacebuilding, postconflict stability and reconstruction, reconciliation, refugee and displaced persons advocacy and support, religious peace traditions, restorative justice, social justice, social movements, sustainable development, transitional justice, trauma healing, and victim advocacy, among many others.

The field is significantly richer because of these and numerous other intellectual and activist traditions, while simultaneously it has been challenged by perspectives like feminism and postcolonial and decolonial theory, which have deepened critical analyses of the field's sometimes limited normative foundations and values.[6] The most famous of the field's limitations, and the subject of numerous essays and books, is referred to as the "liberal peacebuilding" paradigm, which emphasizes expert knowledge and international organizations over local knowledge and community, and teleologically positions free markets and liberal democracy as the universal path toward peace.[7] Many of these enduring dilemmas can be seen, for example, in the contested history of the U.S. government's U.S. Institute of Peace (USIP). Where some see the USIP as a champion of the world's most promising peace practitioners, others see it as an agent of neocolonialism that stifles peace practitioners who do not support the liberal peace paradigm; and where some see it as a counter to the influence of militarism in U.S. foreign policy, others consider it to be a part of the U.S. national security apparatus.[8]

Where there is depth and dynamism, there is also debate. This is especially evident in scholarly, practitioner, and policy disagreements over how to intercede to stop ongoing armed conflicts, which are characterized by a range of assumptions and analytical paradigms that are occasionally in tension with one another. Earlier approaches emphasized conflict regulation,[9] and later conflict management, to underscore the urgent need of containing ongoing violence as the primary and immediate aim of peace negotiators.[10]

Nevertheless, scholars in peace and conflict studies have come to understand that conflict is a normal and necessary aspect of social life—something that philosophers long ago termed "agonism," or ἀγών in ancient Greek, meaning a kind of conflict that is productive and leads to positive, more just, and more peaceful changes to social systems. Karl Marx understood this and considered violent conflict to be a perfectly ethical engine of social change.[11] Jane Addams understood this and argued peaceful, constructive conflict was necessary for social justice.[12] Ella Baker understood this and saw that peaceful struggle was necessary for civil rights. King and Malcolm X understood that conflict was necessary for freedom.[13]

Across peace and conflict studies, approaches that consider conflict as something to be contained or managed have been criticized not only for their relatively narrow focus but also for bracketing out important structural conditions and drivers of violence. As our colleague Kevin Avruch notes, even the names we use to delimit the field carry "deep moral and political assumptions about the nature of people and the world."[14] Terms such as "conflict regulation" and "conflict management" encourage us "to adopt a realist or neorealist position about the nature of conflict and potentials for inducing change," which leads us to assume that the deep causes of a conflict are "beyond our reach, untouchable, located in human nature or the very nature of the conflict system."[15] As a result, those who define the field in terms of regulation or management aim "to achieve balance, stability or deterrence, and not much more."[16]

When the Australian academic and public servant John Burton proposed the term "conflict resolution" in opposition to "regulation" or "management," he did so as a way of grounding an explicit moral critique of the practical implications of the state and power-centered perspectives on conflict that dominated traditional international relations, as well as the settlement-oriented goals of mediation and alternative dispute resolution.[17] To seek to regulate, manage, mediate, or negotiate conflict, in Burton's view, is to ignore the possibility of resolution.

When our colleague John Paul Lederach, in turn, advanced "conflict transformation" as an alternative, he intended it to be an implicit moral critique of what he saw as the limited aspirations of Burton's conception.[18] The goal of conflict transformation, as opposed to resolution, would be not merely to

resolve a conflict but to fundamentally change the social relationships between enemies and build lasting reconciliation, even while accepting certain
forms of conflict and disagreement as elements of a healthy social life.[19] This
would require looking beyond cases of large-scale armed conflict or the role
of elites. We must also examine how violence and conflict may permeate all
levels of society—how it is *experienced*.

In the early 2000s Lisa Schirch, a professor and practitioner, proposed
that the field be known as "peacebuilding," arguing that Lederach's theoretical conception of conflict transformation could be further expanded.[20]
Peacebuilding would involve both structural and systemic changes within a
conflict and entail interventions by various kinds of actors, both external and
domestic, into those conflict settings, most often in postviolence contexts.
Other scholar-practitioners have developed concepts like "justpeace," a term
meant to capture how justice and peace are inextricably linked—and thus
inseparable—requiring that policies and strategies advancing peace be oriented toward a more capacious and integrated understanding of justice.[21]

Moreover, in African American political traditions, pastor and scholar
Beverly Goines reminds us, "peace is an ambiguous term when it comes to
the experience of black people in the United States."[22] While conceptions of
social justice and negative peace have been adopted and used throughout
the field of peace and conflict studies, these concepts were first developed
by thinkers like Martin Luther King Jr., Ida Wells, and Marcus Garvey—
thinkers whose ideas were forged in relationship with social action. By the
turn of the twentieth century, Goines writes, the peace movement in the
United States focused on educating the masses, Christianizing and modern-
izing people around the globe, and promoting the scientific study of international disputes. From the perspective of Black Americans and people of
color, Goines continues, these early formations of the peace movement and
peace and conflict studies were based on a definition of peace that was elitist,
where "the oppressor has to assume that the oppressed are either 'happy' if
the oppressed are not actively making their distress known, or that the oppressed have acquiesced to whatever ideology that is used to legitimize their
inferior human status."[23] This was the "negative peace" that King wrote about
in his "Letter from a Birmingham Jail"—a kind of peace "espoused by white
moderates that prioritizes 'order' over a 'positive peace which is the presence
of justice.'"[24] W. E. B. Du Bois, Goines continues, linked war and peace with
the problems of democracy, while activist Paul Robeson understood that
peace and freedom were connected "and saw the Afro-American people's

quest for social justice as the way to address both issues."[25] Thus, in the tradition of Black thought, one might say that social justice includes peace, where social justice signifies positive peace *plus* freedom. For Black people or any oppressed people, Goines concludes, peace and justice are therefore inseparable—and, in fact, conflict is necessary for achieving social justice.

Taking Goines' critique seriously, we therefore argue that the "peace and conflict" field would benefit enormously from a more thorough theoretical engagement with the notion of agonism, and a more visible practical engagement with generative conflict tactics and strategies as paths toward social justice. Conflict is always a tool, but only sometimes a weapon. It belongs in the peacebuilder's toolkit.

Ethics, Dilemmas, and Wicked Problems

Each of these conceptions about what the field *is* carries explicit assumptions about *what* constitutes practice in that field and *how* to best pursue this work. These assumptions, therefore, bring with them an unavoidable additional ethical critique about the way conflict should be thought about and acted upon. As a consequence, engaging with conflict as a scholar, practitioner, or scholar-practitioner is always a value-laden endeavor. Taking action means taking risks, and perhaps taking sides. Regardless of whether one seeks to regulate, manage, resolve, or transform conflicts, to build peace, or to enhance social justice, one inevitably does this in order to reduce harm and do good.[26] Questions of ethics arise immediately.

In theory, the question of ethics in peace work is straightforward: *ethics involves developing a coherent system of normative principles to examine and inform actions that recognize or advance human dignity and well-being, or that otherwise minimize or end human suffering.* Ethical conduct therefore concerns applying these principles to guide our actions.

In practice, however, the situation is much more complex. Practitioners are often confronted by decisions where there is no obviously correct or best choice of what to do. If the intention is to "do good," whose good counts? How does one prioritize and judge between competing perspectives on the good? And how does one know how to act and when to act? The defining feature of *dilemmas* is that they arise because *dire situations may demand some form of response, but often all of the options may incur undesirable consequences of some form or another.* Dilemmas are not merely ordinary

problems. Whereas problems admit of some solution, dilemmas are consti-
tutively defined by circumstances that allow for only suboptimal outcomes.
Dilemmas are wicked problems, or problems whose *causes and consequences
are so inextricably intertwined that one can't understand them, let alone cope
with them, separately.*[27]

How should peace and conflict studies address wicked ethical problems
that arise in the field? Much of the existing scholarly literature either focuses
on highly abstract questions about agency and responsibility or focuses narrowly on a specific set of circumstances, such as the venerable "just war" tradition, which provides moral guidance on whether and how to fight wars. The
UN's Responsibility to Protect norm, itself firmly rooted in just war thinking,
has widened that discussion somewhat to evaluate under what conditions external military intervention is morally and legally permissible to stop severe
human rights violations. For those readers in the Global North, think back
to Malcolm X's "The Ballot or the Bullet" and consider this question: How
many people in your social circles who tend to support UN interventions in
countries across the Global South would *also* support UN interventions on
behalf of the rights of minorities in their own country? On behalf of French
Algerians in the suburbs of Paris? On behalf of migrants kept in cages, for-
cibly separated from their children, on the southern border of the United
States? On behalf of detainees who remain, for decades, in Guantánamo Bay?

But ethical dilemmas appear in many other dimensions of peace work,
beyond the international realm of conflict management, peacebuilding,
transitional justice, and human rights. The details will vary across cases and
circumstances, but in schematic form we can identify ethical dilemmas as
concerning:

Actors: The presence of many actors with a stake in a conflict can raise numerous challenges. For instance, who should be recognized to speak
for and in the interest of victims, especially where privileging one group
risks sidelining others? Under what circumstances is it morally defensible to work with major agents or enablers of violence, and what are
the trade-offs involved? How should the needs and desires of multiple
victim groups be balanced when they are in tension? Who decides?

Values: When values are in conflict, such as the immediate end of direct
violence versus long-term justice, how does one decide which values
to support? How should values and efficacy of outcome be balanced?
What criteria should one use to justify these decisions?

Motives: It is commonplace to argue that ethical action requires ethical motives, but the latter are often insufficient for success. How important should motives be for assessing the ethical integrity of action? How much weight should be given to ethically sound motives over, say, efficacy of outcome? Conversely, what if the driving motives of actions are not wholly ethically defensible, though the outcome is desirable? How are motivations linked to interests, preferences, values, and goals?

Actions: Even when motivations are defensible, how does one choose among competing options that each may have problematic or otherwise unforeseen consequences? What is the proper balance between reformist approaches, which may be based on piecemeal, steady, and cumulative change, and radical approaches, which may push sharp, disruptive, and intentionally unsettling strategies, even if nonviolent? And when, if ever, is violence a legitimate means of resistance or of bringing about change?

Time frames: How should one balance the needs of the present with the needs of the future, especially when they are at odds or when the long-term consequences of possible actions are unclear? Similarly, how should one balance demands for accountability for past transgressions with the needs of the present or future?

These broad analytical quandaries and categories help frame the complexity of specific dilemmas as they emerge. They also represent a significant theme across contributions to this volume. As Elizabeth Hume and Jessica Baumgardner-Zuzik discuss in their chapter, throughout the field there is relatively little by way of sustained reflection, empirical data, or ethical guidelines that would help one navigate wicked ethical problems. We hasten to note that the field of humanitarianism, which involves using emergency aid to alleviate ongoing severe human suffering, is an exception with an especially rich tradition of ethical reflection. That tradition informs peace work, and it serves as an important starting point for the reflections in this volume.

A core element of modern humanitarianism is found in peace scholar Mary Anderson's articulation of the "Do no harm" principle, which established that *ethical humanitarian action must be judged primarily by its consequences and not only its intentions*; to do no harm is to ensure that an action must not contribute to greater suffering or danger for the targeted civilian population. This principle was animated by concerns that humanitarian aid and relief could, under many conditions, cause or prolong conflicts, which led to more suffering than relief.[28] Thus, for Anderson, "Do no harm" served as a clear answer to a pressing ethical dilemma she and others in the humanitarian and development fields faced in their daily work.

Of course, one should not place people in danger as a consequence of one's actions, and therefore one should think carefully about which activities have the potential to cause harm and how much risk is justified. Yet, as Agnieszka Paczyńska and Susan Hirsch have noted, "do no harm" is often insufficiently precise for informing practice and may prove incapacitating in circumstances where any action carries adverse consequences. Crucially, not acting is itself a form of action—it too carries consequences. In response to the limitations of "Do no harm," Hugo Slim has argued that "the call to do good is a much more positive professional motivation than the more censorious call to avoid doing harm," while Larissa Fast has explored the concomitant dangers aid workers face in carrying out their work.[29]

Humanitarianism largely focuses on the alleviation of current suffering and the protection of civilians in immediate danger. Its time horizon is intentionally limited in scope to the present and near future. However, wicked problems have impacts that extend beyond immediate harms and that arise in many other contexts of peace work, including longer-term peacebuilding. The question of ethical action may be ambiguous where an external third party—such as a peacebuilding or economic development organization—is financed by international organizations and governments that have material interests in promoting peace or social change. Because peacebuilding "entails the most intensive and wide-ranging intervention by others into the conflict system (society or culture),"[30] peacebuilding is among the most ethically fraught areas of practice in the peace and conflict studies field.

Especially problematic is the occasional unwillingness of third-party peacebuilders to reflect on the connection between their own moral values and the values, goals, and interests of the governments, organizations, and institutions for which they work.[31] The prevailing institutional arrangements across the world—i.e., which donor countries and global NGOs mobilize in service of peacebuilding work—were largely created in reference to the economic and political arrangements of nation-states and therefore "'morally' exclude sub-national groups from articulating their ethical claims against nation-states in [an] internationally and legally sanctioned forum."[32] Peacebuilding work, no matter how self-consciously and ethically designed, often points normatively toward the goal of state-building, which in turn may entail the suppression of claims made by subnational groups that view the state itself, with its monopoly over the instruments of violence, as a party in the conflict.[33]

Indeed, from this vantage point, the principle "Do no harm" is a low bar for ethical standards of practice. In such contexts, Paczyńska and Hirsch remind us, peacebuilders might see themselves as doing good, and certainly not doing harm, yet be unable to reflect upon the content of this good and how it might not be serving the interests of the people they are attempting to help.

Writing from her own experience as a professional working for international organizations, scholar and peacebuilder Reina Neufeldt has proposed an "action-reflection" model to help peacebuilding practitioners think clearly and ethically about the implications of their good interventions.[34] Critical self-reflection along these lines is also necessary within peacebuilding organizations, especially if these organizations want to avoid falling into patterns that have been shown to limit individual members' ability to reflect ethically. This may include groupthink scenarios or a focus on procedures and process over substantive outcomes.[35] Without a mechanism for such action-reflection, Neufeldt argues, it becomes exceedingly difficult for the peacebuilder to avoid co-optation, to balance donor demands with local integrity, or to decide how to formulate, prioritize, and change goals in shifting contexts.

For Tim Murithi, the ethical practice of peacebuilding must also admit the moral validity of actors who are not considered worthy of having their voices heard within state-centric institutions, even when this means morally validating those who use violence as a means of addressing their interests in a conflict.[36] From Murithi's perspective, ethical reflection requires peacebuilders to include all human groups in the moral vision of a peace process, without demanding that individuals give up their group affiliation as a prerequisite for moral standing. The issue of ethics and leadership and the enduring dilemmas that arise in the practice of peacebuilding, Murithi argues, center around dilemmas like navigating trade-offs between accepting violent actors and denouncing violence; operationalizing positive peace without compromising the principles of justice and democratic inclusion; accepting victims' experiences without allowing victims to "other" their former perpetrators; and balancing competing claims for justice with an understanding that it is often the marginalized and the weak within the power structure of society who are problematically expected to do the forgiving and reconciling. Pamina Firchow's important work on "everyday peace" also captures the pressing need for returning authority and agency to the communities most affected by violence rather than assuming a univalent, universal understanding of peace can be implemented in any context.[37]

Themes and Dilemmas

With this book we have set out to dramatically expand the conversation about ethical dilemmas. We have done this by including more voices from a wider range of change-oriented action than is usually considered to be part of the peace and conflict studies field. Our reasoning is simple: the field is larger than many recognize. This is clear, if provocative: inequality and racism in the United States and the Global North are peace and conflict studies issues; debate over the legitimacy of violence, anywhere in the world, is a peace and justice issue; debates over policing and the state, anywhere in the world, are peace and justice issues; collective action and contentious politics, anywhere in the world, are peace and justice issues. It is important to ask ethics-oriented questions about the outcomes of foreign interventions, of course. It is crucial that we develop tools that build a better world for everyone rather than hectoring and "helping" others elsewhere. In talking about ethics of action, then, the lessons are for all of us and wherever we work for change.

Section I: Violence

The book's first section returns to the wicked problem that started this chapter: the role of violence. There is a broad and long-standing consensus that nonviolence is the most ethical approach to social change, an argument built off the pioneering thought and action of Mahatma Gandhi and Martin Luther King, Jr. This deep consensus about ethics has recently been joined by a compelling argument for the efficacy of nonviolence. Political scientists Erica Chenoweth and Maria Stephan have drawn on an impressive array of data to argue that nonviolence is, on average, a more effective tactic for social change.[38] Say what you will about ethics, they argue, strategic nonviolence is more likely to work. Their work is as groundbreaking as it is compelling, requiring a fresh round of theorizing about the role of violence as tactic and strategy. Indeed, social action in the wake of this ethical-tactical consensus raises new wicked problems.

Tony Gaskew, writing as both a scholar and a former police officer, steps into this conversation with a provocative argument: African Americans have little choice but to use violence to protect themselves against systematic state repression. Gaskew's argument for armed self-defense explicitly takes on the tradition of Black nonviolent resistance as ineffective and too accommodating, noting that armed resistance has long been an important feature of the civil rights movement. This contribution covers a range of ethical dilemmas in the American context and provides an important—if potentially uncomfortable—statement on responding to injustice. His call to arms, as it were, is all the more salient in the current American context. We write in the midst of a broad-based and nonviolent radical justice movement, in the form of Black Lives Matter. We also write in the midst of a strident and violent response, in the form of widespread tear-gassing, detainments, and beatings by the police and federal law enforcement, violent assaults and shootings by armed right-wing vigilante groups, and numerous racist attacks.

Writing from the heart of the contemporary Poor People's Campaign, Liz Theoharis and Noam Sandweiss-Back argue that America's recent history is "bloodied with examples" of how government positions create unlivable conditions for the poor in the United States, the richest country in the world, that are nothing less than "violence of the highest order." From the Great Recession of 2008, "the very speculators and financial institutions that were responsible for the devastating crash were bailed out" by the Obama administration, and twelve years later, in March 2020, "the Federal Reserve materialized over a trillion dollars to buoy Wall Street in the days immediately following the COVID-19 shutdown." Meanwhile, the U.S. government offered "threadbare and temporary support to most citizens and completely lock[ed] out millions in undocumented communities, homeless encampments, prisons, and more." The U.S. death toll from the COVID-19 pandemic grew to nearly 900,000 while this book was under review. Poor, minority, and marginalized communities have suffered the most. Notably, the suffering comes in two forms, from the disease itself and from a shredded social safety net, seemingly designed to bail out only the wealthiest. Nonviolent mass social awakening, Theoharis and Sandweiss-Back insist, is the only way these systems of violence can be undone.

Nonviolence raises its own ethical dilemmas, political scientist Kirssa Cline Ryckman suggests: while there is no ethical justification for violence, unchecked state repression may put additional activist lives at risk. Violence, Ryckman argues, should not be a proactive strategy, but may be a tactic of last resort. This approach has the effect of pitting two ethical positions against one another: the tactical and Gandhian commitment to nonviolence advanced by Chenoweth and Theoharis and Sandweiss-Back on one side and the fundamental commitment to the survival and self-preservation advanced by Gaskew on the other.

Ashley Bohrer, a philosopher with a long history of political activism, completes the section with an important set of reflections on the trade-offs between reformist efforts to work within the political system and revolutionary pursuit of deep transformational change through a more confrontational posture, even while remaining peaceful. Bohrer moves beyond the standard academic (and often reductive) frame of "reform or transform" to explore how both approaches can be used strategically, depending on circumstances, while also noting the inherent tensions that this draws out.

What this section makes clear, yet again, is the tension baked into the dual pursuit of peace and justice. What are the practical implications of tactical nonviolence and of self-defense? How shall we think about the interplay between these two commitments? What impact does the theory have in shaping personal action and social movement strategy? These are important questions, and getting the answers right is crucial for both means and ends.

Section II: Leadership and Organizations

From a specific tactic, our attention shifts to the role of leadership and alliances in ethical action. For Minh Dang, ethical representation as a leader is especially fraught in movements that seek to integrate the voices of those most affected by injustice. A survivor of human trafficking and now head of the UK-based antitrafficking network Survivor Alliance, Dang lays out the challenges of building a justice movement among survivors who have myriad economic and political needs and demands, while also resisting dehumanization, "pedestalization," and tokenization from the wider public and media. Dang concludes with steps for concrete, tactical alliances, support for lived experience expertise, and sustained pressure for the reform of the laws and practices that facilitate human trafficking. While acknowledging that these wicked problems might not be completely overcome, her discussion of the practical ways in which leaders can ameliorate them resonates with Bohrer's chapter on moving beyond simple revolution/reform binaries.

University president and social movement scholar Daniel Myers considers the role of allies in leadership positions, offering his personal history of leadership activism as an ally to the movement for LGBTQ rights. Myers identifies a number of important roles that allies can play, such as cultivating support from broader audiences, but he also underscores a host of knotty challenges, including those related to identity, representation, authenticity, and voice, asking, in essence, *Who ought to speak for whom?*

Scholars have long known that a successful movement requires, among other things, adequate material resources to survive setbacks and to have a long-term impact. alicia sanchez gill, an organizer with the Black radical feminist network INCITE! and now a philanthropy activist, examines how funding from influential donors can play a pivotal role in activism while also introducing risks, including the co-optation of the mission, the sidelining of experiential knowledge and activist voices, and the dilution of the core message. Her chapter is enriched with contributions provided by other activist leaders, including Darakshan Raja of the Justice for Muslims Collective, Janis Rosheuvel with Solidaire, Donald Anthonyson with Families for Freedom, Zuri C. Tau of Liberatory Research and Social Insights Research, Latishia James-Portis, a chaplain and peace facilitator, and Travis Akil-Brookes and Nicole Newman, organizers in Washington, D.C. Each shares a grassroots perspective on profound ethical dilemmas that leadership presents in community organizing, direct action, and social movement work.

Longtime conflict resolution facilitator and now scholar Philip Gamaghelyan explores another angle to the challenges of leadership and activism. Drawing on the work of science fiction authors Ursula K. Le Guin and N. K. Jemisin, who have long engaged with justice issues in their work, Gamaghelyan identifies a range of political actors: those who acquiesce to injustice, those who leave, and the ones who "stay and fight." The latter group often faces significant challenges, including the dilemma of navigating both domestic pressures from a repressive state and well-meaning but overwhelming influence from outside allies. Gamaghelyan uses his experiences from across the South Caucasus to explore the ethical trade-offs in alliance building in these complex situations.

Reina Neufeldt, a peacebuilder and scholar, explores how individual unwillingness to examine core values can provide unmerited justification for certain policies and prevent important and thoughtful deliberation with affected parties. She concludes with several recommendations for promoting ethical reflection and humility in peacebuilding, which in turn can democratize leadership opportunities.

Section III: Systems and Institutions

Powerful institutions shape the peacebuilding terrain. They act as catalysts for change, and they also create obstacles and challenges. Peacebuilding institutions themselves create and sustain dilemmas. The book's third and final section highlights a few of these dilemmas through a number of carefully selected case studies.

Deena R. Hurwitz directly engages the lived-experience dilemmas that emerge from working alongside vulnerable communities. Hurwitz's work on human rights advocates and lawyers—who work in places where the rule of law is compromised, whether in a country ruled by a dictator or an ostensibly liberal democracy governed by national security imperatives or military law—presents a number of ethical dilemmas that arise when advocates encounter the manipulation of law for political ends. These are "dilemmas of personal conscience and of professional responsibility, of moral accountability and complicity—and how that is defined, by and for whom," she writes. What happens when a vulnerable local partner asks a human rights lawyer or advocate to carry sensitive documents, letters, notes, or photographs? How forthright should one be with the security forces and border personnel of a rights-abusing state? Should one answer truthfully as a matter of principle, although it may compromise the safety of colleagues, clients, friends? Should one respond with false information? Or should one be evasive or give partial information? Hurwitz's questions strike at the core of two major themes of this book so far, as those seeking social change balance the well-being of the vulnerable people they hope to serve with the realities of state violence and the constraints of oppressive institutions.

George Lopez, a scholar and former member of the UN expert panel on sanctions on North Korea, and Beatrix Geaghan-Breiner, a student of international relations, explore the normative issues surrounding the contemporary uses of economic sanctions in international politics, which raise complex ethical questions because of their often profound economic impact on civilians. In principle, sanctions are used to compel repressive states to change their violent behavior, but as the authors note, shorn of specific and clear evaluative criteria and benchmarks, sanctions can exacerbate civilian suffering. At its most basic, the wicked problem is whether it is worth inflicting new harms in order to end ongoing harms. Lopez and Geaghan-Breiner use the case of U.S. sanctions against Venezuela to develop a norm of the "responsibility to restore" targeted economies in critical areas. They conclude by outlining the various practitioner and scholarly tasks needed to bring this to fruition.

Ernesto Verdeja turns to the wicked problems at the heart of international mass atrocity prevention work, today characterized by a complex assortment of human rights NGOs, governments, and regional actors like the UN, African Union (AU), EU, and many other institutions. Drawing on interviews with practitioners, prior scholarship, and his own experiences in prevention practice, Verdeja sketches a general ethical framework for atrocity prevention. This exercise highlights key ethical dilemmas, including challenges concerning representation (who speaks for the vulnerable?), practical issues around the use of force to protect civilians, and the question of compromising with perpetrators in order to stop massacres. Verdeja concludes with a set of general reflective guidelines that help limit the scope of these dilemmas, even if they cannot be eliminated.

Mediation scholar Laurie Nathan's contribution extends this discussion, focusing on dilemmas that arise in mediating armed conflict between violent states, insurgents, and external actors like the UN and the AU. Nathan is a scholar, as well as a longtime conflict mediator with the UN and AU, and here he examines how mediation efforts often require compromises with armed groups that can leave accountability and justice demands off the table. Exploring AU mediation efforts in Côte d'Ivoire (2010–2011), Libya (2011) and Syria (2011–2018), Nathan characterizes the trade-offs between securing immediate peace and protection of civilians and the long-term pursuit of justice as fundamentally a "situational incompatibility of good norms," where a hard choice is required among norms that are both desirable and, in different circumstances, generally compatible.

For example, even if peacebuilders can identify criteria for deciding when to respond to violence against civilians, there is still the question of how to respond. Felicity Gray, a scholar of international peacekeeping, notes that most civilian protection programs—including those of the UN and AU—rely on the use of armed peacekeepers, based on the logic that a credible threat of retaliation will dissuade would-be perpetrators from harming civilians. Gray identifies limitations with this status quo and offers instead a *nonviolent civilian protection approach* that is rooted in a wide range of nonviolence practices and strategies. The goal is to reduce overall levels of harm and threat. This more extensive and integrated approach allows Gray to explore the alternative set of dilemmas that may emerge were this approach to be adopted. It is a provocative and thoughtful chapter that, by essentially putting the Responsibility to Protect literature into conversation with the strategic nonviolence literature, goes against the dominant trend of international civilian protection practice.

The contribution of South Africa–based scholar and practitioner Tim Murithi explores the ethical dilemmas surrounding transitional justice. Transitional justice itself might be best thought of as a wicked problem, since it involves thinking about how societies moving away from authoritarianism or armed conflict to peace and democratic rule reckon with legacies of violence and impunity. Murithi's extensive work in the South African transition and across Africa provides him with a vantage point to examine profound problems in a number of contexts. These include questions of how to balance possibly divisive demands for accountability and economic redistribution with calls for societal reconciliation, which may happen at the expense of victims and historically marginalized groups. Murithi's chapter notes how these dilemmas become especially acute when incoming political leaders are constrained in their ability to address the past. A failure to look back may limit the use of institutions like truth and reconciliation commissions to engage with issues of justice, reconciliation, and trauma.

We conclude this section with observations on the dilemmas arising in two institutions that represent where most of our authors and many of our readers are based: NGOs and the academy. Elizabeth Hume and Jessica Baumgardner-Zuzik of the Alliance for Peacebuilding (AfP) rightly note the need for rigorous research in crafting viable and effective peacebuilding programs. The AfP is a central player in modern peacebuilding, with over 120 member organizations from across the humanitarian aid and development worlds, as well major academic research institutions. Hume, AfP's executive director, and Baumgardner-Zuzik, the senior director of learning and evaluation at AfP, are especially concerned with research that privileges technical mastery and methodological sophistication while ignoring issues of context, the rights and moral standing of research subjects (including vulnerable civilians), and asymmetric power relations. Their chapter lays out a set of practical guidelines for ethically informed research that underscores the need for conflict sensitivity, protections for vulnerable research populations, and clearer standards for program design, implementation, and monitoring.

Finally, the academy. Typically, discussions about ethical action in peacebuilding focus on questions of policy choice and implementation, mobilization tactics and strategies, and research ethics for program and policy development. However, the education of future peacebuilders is itself a critical ethical decision point, one that may have long-term effects on an individual's understanding of and conduct in the world. Peace and conflict scholars Agnieszka Paczyńska and Susan F. Hirsch bring these issues to the fore in their examination of the ethical dilemmas surrounding teaching about violence and peace in conflict zones, whether in our own communities or in faraway locales. Drawing on the increasingly common use of experiential learning and field-based courses, Paczyńska and Hirsch discuss how students with strong normative commitments but limited understanding of the world find their assumptions challenged and questioned in the field, where problems and possible solutions are much more complex than they may have expected. The authors highlight a host of wicked problems and provide concrete suggestions on how to make field courses more ethically reflective, noting the importance of consent, inclusivity, and listening to local voices. The result is an approach that underscores the value of field experience while also encouraging humility and awareness of the claims and agency of local actors. These dilemmas and responses are relevant well beyond the academy, as they apply to any field-based efforts, whether undertaken by researchers or activists.

Crosscutting Themes: Identity and Inequality

Reading across these three sections suggests that identity and inequality are recurring. It is clear to any contemporary reader that inequality and identity are critical rallying points for dynamic and emergent social action. We would like to take a moment here to further argue that these are two areas where the peace and conflict field simply must catch up with activity on the ground. As male editors in the Global North, and with our own positionalities and identities, we emphasize intersectional issues related to the ethics of peace, rights, and justice work that occur in the context of strategic nonviolence, activism, and contentious politics, in the dilemmas of leadership and organizations, and in efforts to change systems and institutions. A number of essays prompt reflection on gender and the politics of sexuality, while taking the principles of intersectionality as a given. We might say there are any number of positionalities at play in the book, as these essays take up questions of gender roles in movements and gender conflicts through issues of race, class, culture, ethnicity, sexuality, and many others.

Nathan, Murithi, Hume and Baumgardner-Zuzik, and Paczyńska and Hirsch thread issues of gender mainstreaming into their discussions of the ethics of movement leadership, peacebuilding, conflict resolution, and critical pedagogies. And it is not for nothing that Theoharis and Sandweiss-Back frame their discussion of the poor people's movement through the words not of the oft-quoted Martin Luther King Jr. but of Coretta Scott King, whose unique positionality as a Black woman and a public figure married to a civil rights icon is deeply entwined with the emergence of her own distinct, but often overlooked, political philosophy.

For Bohrer—whose chapter sets out not to chart a set of gendered ethical dilemmas in social justice movement work but rather to highlight competing options for movement activists weighing prefigurative and harm-reduction approaches to social movement work—the path to a politics that can galvanize movements into seeing the radical possibilities of a more just and liberated world is paved by the contrasting ideas of Audre Lorde and Serene Khader. Where Lorde provides a philosophy that helps activists avoid replicating oppressive structures, Khader suggests resistance strategies should be assessed on their efficacy. From these two contrasting visions of feminist theory, Bohrer extracts a set of provocative and invaluable questions that movement leaders must ask themselves about institutions that have largely been built by and for men from dominant classes, castes, ethnicities, and religions.

Scholar and organizer Dang focuses directly on the gendered dilemmas faced by movement leaders. Drawing on her experience leading one of the few survivor-led NGOs dedicated to empowering survivors of slavery and human trafficking, Dang argues that lived experience expertise is essential for social movements, but that survivors face a "double bind." How should survivors of sexual or gender violence bring their experience to the movement when this leadership and advocacy exposes the individual to the male-centric social gaze that permeates wider society and that constantly generates narratives that reduce the agency of survivor-activists to the "gory details" of their past experiences? A second dilemma, Dang argues, emerges if we are to take seriously standpoint epistemology— the fact that what we see and know is based on who we are and where we stand. The challenge is clear: how to cultivate crucial movement allies from communities that lack the unique perspective that comes from lived experience of human trafficking. These dilemmas also emerge *within* survivor movements, she writes, making it all the more important for cisgendered women like herself to explicitly look for opportunities to empathize and educate themselves about the experiences of survivors who are nonbinary or transgendered women and men.

Myers picks up the question of gender identities and the perceived legitimacy of allies in leadership roles of organizations. Reflecting on his experiences as a lifelong activist for LGBTQ rights, Myers unpacks his time as an "ally in leadership" for Queer Nation during his years as a student, and how these experiences helped him understand how to be a better ally. Legitimacy in social movements depends heavily on endorsement from the community, Myers writes, and such endorsements depend, in turn, on identity markers. In the context of LGBTQ rights movements, these identity markers are necessarily gendered. Because the Queer Nation movement needed people to claim their identities as queer folks, Myers writes, having straight leaders with gender markers of privilege and advantage would have undermined the purpose of a social movement designed to empower and defend the rights of those whose gender identity markers rendered them marginalized and oppressed within a wider society.

Drawing on years of experience as a leader and organizer with INCITE!, a network of radical feminists of color, gill picks up the conversation around the gendered dilemmas of leadership in movement spaces. gill invites a series of essays, in the style of self-reflective narratives, which elevate the stories of leaders who live at the intersections of many marginalized identities. In the crucible of their struggles, gill presents the internal conflicts faced by movement leaders with intersectional identities, as they are confronted with competing needs and demands from the many marginalized groups with whom they identify.

Darakshan Raja's contribution to gill's chapter presents the challenges experienced by Muslim women in leftist spaces who face struggles that white and cis-gendered male leaders don't encounter when confronting state violence, gendered violence, harassment, and the surveillance state that targets Black and Brown Muslim women and femmes.

In their essay, Janis Rosheuvel and Donald Anthonyson unpack many of the inherently gendered dynamics they encounter in the Families for Freedom movement, which was founded to resist the brutality of the mass deportation system that emerged following passage of the "1996 Laws" and the post-9/11 criminalization of migrant life, and unites members from Black, Asian, Latinx, undocumented, and documented communities.

gill also presents the reflections of Zuri C. Tau, who was spurred to help found an organization, Building Local Organizing for Community Safety, after Atlanta police burst into the home of a grandmother and shot her. For Tau, the burden of this memory—of an old woman, sitting in a neat and clean kitchen, in dignity before a most undignified and inhuman death—has led to a lifetime of doubts and regrets, all of them entangled in questions of positionality and identity: *Did I give up too soon? Did our work mean less because those who were most vulnerable were not the most visible? Were we the right organizers for the fight?*

In her essay Latishia James-Portis, a movement chaplain and facilitator, reverses the lens, looking at regrets and dilemmas not from the perspective of the advocate but from the perspective of movement leaders who are traumatized. What should be done when it is that trauma which sustains involvement in the movement, while simultaneously preventing one from showing up as the person they claim to be?

Nicole Newman's narrative of organizing in Washington, D.C., explores the exhausting toll exacted on individuals when "white folks in organizing and advocacy spaces suddenly see the validity of lived experience and then want to hear from people impacted, though many Black people have been restating the same needs, root causes, and interventions for years." The transactional and performative experiences that many white organizers and white advocates ask of their Black allies require a kind of emotional labor that is itself toxic and exploitative, while entailing a "constant flattening of our experiences and a belief that there is one way to be Black and working class"— which is all too familiar to anyone exposed to the way white narratives have long attempted to make sense of Black lives.

Finally, picking up on these themes of translations, transactions, and performances, gill presents organizer Travis Akil Brookes's reflections on movement masculinities. "Masculinity, masculinity, masculinity. Shit's toxic. I can read, I can go to the workshops, I can have the conversations," Brookes recounts, "but if I'm not giving myself the grace to unlearn masculinity then I'm more of a problem than a 'solution.'" In Brookes's critical reflection of the dynamic between the construction and deconstruction of his masculine self-identity as a movement leader, we find a hard-won awareness that "it was the labor of Black, trans, and queer folk that helped me to see that where I am now, where I was then, and where I will be is all a part of the journey. And if

I'm really about unlearning this shit, then I gotta recognize that I'm going to fall down, I'm going to fail, that I need to center healing, and that I should not do any of this alone—independence is a deeply entrenched belief of masculinity and white supremacy."

Brookes's lesson that the myth of independence is an entrenched belief in both masculinity and white supremacy reverberates through Gaskew's chapter. In many ways, we can read Gaskew's contribution to this volume as a kind of coming-of-age story. The profound awareness that his life's purpose is to wage war on white supremacy is mediated by an intergenerational passing of the torch from father to son, as Gaskew frames the Black Radical Tradition as both an intergenerational birthright and a spiritual inheritance. To be clear, the Black Radical Tradition is not defined or limited by constructs of gender. Gaskew is not suggesting that it is. Importantly, in his vision this tradition has always been a collective voice for liberation. Contrast this with Brookes's observation that the myth of independence is entrenched in social constructs of masculinity and white supremacy.

The myth that Europeans and Euro-Americans (and "white" people more generally) are a self-sufficient stock, who exalt individual independence as a moral virtue, conceals the fact that the material wealth and concrete power of European and Euro-American societies have been built upon the oppression, exploitation, enslavement, and genocide of tens of millions of fellow human beings.[39] The myth of the self-sufficient white man is epitomized by the idealized image of the frontiersman in the American imagination, but the belief that this ideal white Euro-American man provides for himself and his family through his own hard work with no help from others conceals the core reality that wealth and "independence" depended very much upon land, labor, resources, and life taken from people of color.[40]

Brookes's reflections on his own journey to embrace a wider and more "connected-to-others" gender identity exposes the myth of individualist masculinism as a subtle but pernicious tool of white supremacy. Gaskew's chapter helps us parse the possible responses to the wicked problems Brookes brings up, as both chapters present the reader with blueprints for the abolition of white supremacy that begin with an understanding that one's self is spiritually connected to others. The transitional moment from childhood that Gaskew recalls is, quite explicitly, a rejection of the ideology (and myth) that independence is the measure of a person. In fact, by presenting the essence of the Black Radical Tradition as an intergenerational birthright and a

spiritual inheritance, Gaskew embraces the principle of *inter*dependence—
not *in*dependence—as a core moral virtue and a powerful means for
liberation.

Wicked problems are present in our philosophies and in our actions,
in efforts to change our own communities and to serve strangers in other
places, in our strategies and in our tactics, in our budgets, organizational
charts, leadership approaches, alliances, and collaborations, and even in our
pronouns—who is the "our" we are referring to?

In order to build this volume, we have done our best to highlight a broad
range of voices from around the world and from various justice struggles.
Some provide more analytically systematic assessments of wicked problems,
while others reflect on these big issues with their lived experiences as the
starting point. It is our hope that this ecumenical approach brings greater nu-
ance and complexity to assessing these ethical dilemmas. Above all, we hope
it expands our ability to take more ethical action.

Where Do We Go from Here?

Wicked Problems provides a number of perspectives that illuminate the extent to which peacebuilding exposes practitioners to complexity. Creativity
and critical thinking have never been more important in the peacebuilding
space, especially with the growing realization that the world's problems, and
solutions to the world's problems, are not straightforward, nor is the world
divided up into clear binaries of good actors and bad actors, nor domestic
and international issues.

Wicked Problems highlights efforts to go beyond narrow academic debates
to examine the practical opportunities and challenges raised by decision-
making dilemmas as they are experienced on the ground. This book can advance scholarship on ethics in a number of academic subfields: international
and global studies, peace and conflict studies, justice studies, and human
rights. Our larger goal, however, is to enhance feedback between practice and
scholarship so practice can inform theory and theory can inform the work of
practitioners, advocates, and peacebuilders.

Our second objective is to broaden the tent of who we consider fellow-
travelers on the justpeace path, as it were. We believe there is room for anyone
committed to the struggle for human rights, social justice, and positive peace,

the last broadly understood as creating sustainable conditions for human flour-ishing. In fact, one of the reasons we decided to build this book is that, despite our different disciplinary backgrounds—sociology, global affairs, political sci-ence—we recognize many similar wicked problems involved in actually *doing* the good work our various academic and movement backgrounds advocate for.

We have a final, and longer-term, objective with this book. Fields of schol-arship and practice such as ours are doomed to fail if people around the world begin to think of us as hypocrites, uncritical and naïve do-gooders, or true believers who objectify others and treat individuals unjustly in pursuit of goals we deem morally good. On occasion, people who work for positive social change—whether they are trying to build bridges, pioneer interfaith work in their city, or fight for rights halfway around the world—justify their actions according to the crudest of ends-oriented utilitarian logics. We often consider the dictum "Do no harm" to be the highest of ethical standards, a North Star of sorts. Yet even here we are confronted with a dilemma. Ethicists from almost every tradition and around the world would consider this to be the least you can do, the bare minimum, the smallest of starting points. Meanwhile, activists like King and Gandhi took actions to put individuals into harm's way for a greater good. Which takes the ethical precedent: the in-dividual or the collective? Your own group members or followers, or others? Oppressors or the oppressed?

Wicked problems indeed.

Whether we are conducting trainings and workshops or on the front lines of peacebuilding, social movements, or other efforts to support social change, we tend to think we're doing good simply because we've shown up and are on the right side of a dichotomy. Reading across this volume helps us to see that it isn't so simple. Avoiding such pitfalls requires honest and col-lective self-reflection. This self-reflection, in turn, requires nuanced ethical conversations. We hope this book is a step in the right direction.

Notes

1. Martin Luther King, Jr., "Beyond Vietnam," in *A Time to Break the Silence: The Essential Works of Martin Luther King, Jr.*, ed., Walter Dean Myers (Boston: Beacon Press, 2013), 78.
2. Malcolm X, "The Ballot or the Bullet," in *Malcolm X Speaks: Selected Speeches and Statements*, ed. George Breitman, (New York: Grove), 23.

3. See generally Stephen Eric Bronner, *Ideas in Action: Political Tradition in the Twentieth Century* (Lanham, MD: Rowman & Littlefield, 2000).

4. See Kevin Avruch, "Does Our Field Have a Centre?," *International Journal of Conflict Engagement and Resolution* 1, no. 1 (2013): 10–31.

5. Avruch, "Does Our Field Have a Centre?"; Louis Kriesberg, "The Development of the Conflict Resolution Field," in *Peacemaking in International Conflict: Methods and Techniques*, ed. I. William Zartman and J. Lewis Rasmussen (Washington D.C.: U.S. Institute of Peace Press, 1997), 51–77.

6. Victoria Fontan, *Descolonización de la Paz* (Cali: Pontificia Universidad Sello Editorial Javeriano, 2013); Kevin Kester, Michalinos Zembylas, Loughlin Sweeney, Kris Hyesoo Lee, Soonjung Kwon, and Jeongim Kwon, "Reflections on Decolonizing Peace Education in Korea: A Critique and Some Decolonial Pedagogic Strategies," *Teaching in Higher Education*, no. 2 (2019), doi: 10.1080/13562517.2019.1644618; Atalia Omer, "Decolonizing Religion and the Practice of Peace: Two Case Studies from the Postcolonial World," *Critical Research on Religion*, no. 3 (2020), doi:10.1177/2050303220924111.

7. Shahrbanou Tadjbakhsh, ed., *Rethinking the Liberal Peace: External Models and Local Alternatives* (London: Routledge, 2011); Roger Mac Ginty, *International Peacebuilding and Local Resistance: Hybrid Forms of Peace* (London: Palgrave, 2011); Edward Newman, Roland Paris, and Oliver Richmond, eds., *New Perspectives on Liberal Peacebuilding* (Tokyo: United Nations University Press, 2009).

8. Michael D. English, *The US Institute of Peace: A Critical History* (Boulder, CO: Lynne Rienner, 2018).

9. Paul Wehr, *Conflict Regulation* (Boulder, CO: Westview Press, 1979).

10. Dennis Sandole and Ingrid Sandole-Staroste, eds., *Conflict Management and Problem Solving: Interpersonal to International Applications* (New York: New York University Press, 1987).

11. Lewis A. Coser, *The Functions of Social Conflict* (London: Routledge, 1956).

12. Jane Addams, *Newer Ideals of Peace* (London: Macmillan, 1906).

13. Peniel E. Joseph, *The Sword and the Shield: The Revolutionary Lives of Malcolm X and Martin Luther King Jr.* (New York: Basic Books, 2020).

14. Avruch, "Does Our Field Have a Centre?," 11.

15. Avruch, "Does Our Field Have a Centre?," 11.

16. Avruch, "Does Our Field Have a Centre?," 11.

17. John Burton, *Conflict: Resolution and Prevention* (New York: St. Martin's Press, 1990).

18. John Paul Lederach, *Preparing for Peace: Conflict Transformation across Cultures* (Syracuse, NY: Syracuse University Press, 1995).

19. Avruch, "Does Our Field Have a Centre?," 11.

20. Lisa Schirch, *Ritual and Symbol in Peacebuilding* (Bloomfield, CT: Kumarian Press, 2005). For a defense of the traditional liberal peace, see David Little, "Religion, Peace and the Origins of Nationalism," in *The Oxford Handbook of Religion, Conflict, and Peacebuilding*, ed. R. Scott Appleby, Atalia Omer, and David Little (Oxford: Oxford University Press, 2015), 70–104.

21. Mac Ginty, *International Peacebuilding and Local Resistance;* Oliver Richmond, *The Transformation of Peace: Rethinking Peace and Conflict Studies* (Basingstoke: Palgrave Macmillan, 2007).

22. Beverly Janet Goines, "Social Justice as Peacebuilding in Black Churches: Where Do We Go from Here?," in *Building Peace in America,* ed. Emily Sample and Douglas Irvin-Erickson (Lanham, MD: Rowman and Littlefield, 2020).

23. Goines, "Social Justice as Peacebuilding in Black Churches," 23.

24. Goines, "Social Justice as Peacebuilding in Black Churches," 23.

25. Goines, "Social Justice as Peacebuilding in Black Churches," 29.

26. Reina C. Neufeldt, *Ethics for Peacebuilders: A Practical Guide* (Lanham, MD: Rowman and Littlefield, 2016).

27. Charles Hauss, *Security 2.0: Dealing with Global Wicked Problems* (Lanham, MD: Rowman & Littlefield, 2015).

28. Mary B. Anderson, *Do No Harm: How Aid Can Support Peace—or War* (Boulder, CO: Lynne Rienner, 1999).

29. Hugo Slim, *Humanitarian Ethics: A Guide to the Morality of Aid in War and Disaster* (Oxford: Oxford University Press, 2015), 7; Larissa Fast, *Aid in Danger: The Perils and Promise of Humanitarianism* (Philadelphia: University of Pennsylvania Press, 2014).

30. Avruch, "Does Our Field Have a Centre?," 11.

31. Neufeldt, *Ethics for Peacebuilders.*

32. Tim Murithi, *The Ethics of Peacebuilding* (Edinburgh: Edinburgh University Press, 2009), 43–70, at 47.

33. Murithi, *Ethics of Peacebuilding,* 161.

34. Neufeldt, *Ethics for Peacebuilders,* 126–128.

35. Neufeldt, *Ethics for Peacebuilders,* 126–128.

36. Murithi, *Ethics of Peacebuilding,* 165.

37. Pamina Firchow, *Reclaiming Everyday Peace: Local Voices in Measurement and Evaluation after War* (Cambridge: Cambridge University Press, 2018).

38. Erica Chenoweth and Maria Stephan, *Why Civil Resistance Works: The Strategic Logic of Nonviolent Conflict* (New York: Columbia University Press, 2011).

39. Michel-Rolph Trouillot, *Silencing the Past: Power and the Production of History* (Boston: Beacon Press, 1995).

40. Roxanne Dunbar-Ortiz, *An Indigenous Peoples' History of the United States* (Boston: Beacon Press, 2014). And see Roxanne Dunbar-Ortiz, *Loaded: A Disarming History of the Second Amendment* (San Francisco: City Lights, 2018).

SECTION I
VIOLENCE

In this section, authors confront one of the most profound dilemmas for social change: What role, if any, should violence play? Nonviolence has long been central to transformative politics and struggles against injustice, whether grounded in secular or religious principles of inherent human dignity that condemn harming others or in strategic arguments that it is comparatively more effective in the long run. Gandhi, Ella Baker, Martin Luther King, Jr., and Jeanne Cordova have all famously argued for the superiority of nonviolent resistance to injustice, and contemporary activists like Yolanda Becerra Vega, Leymah Gbowee, and Liz Theoharis have continued the tradition of peaceful but disruptive resistance. Research seems to confirm its efficacy: Erica Chenoweth and Maria Stephan have shown that nonviolent movements tend to be more successful than violent ones. But others, like Frantz Fanon and Malcolm X (at one point), saw violence as a key element—even a central part—of transformative struggle, especially when other methods seem to fail.

We begin with contributions that take starkly different positions on this question—indeed, some contributors have a strong response to the legitimacy of violence, even while reflecting on the challenges and consequences of their positions. The discussion then expands to a related but distinct ethical dilemma: When should one focus on harm reduction and work within systems of injustice, and when should a prefigurative approach rejecting compromise in either means or ends be adopted, one that rejects the system in toto? This final issue is central to much of practical peacebuilding work and receives thoughtful consideration in the following pages.

1

The Ritual of Black Armed Resistance

Police Abolition through the Eyes of the Black Radical Tradition

Tony Gaskew

In 1973, as a nine-year-old, I was standing on the corner of 117th and Michigan Avenue, which is on the far Southside of Chicago in the Roseland neighborhood known today as the "Wild 100s," waiting on a CTA bus to go home after school.[1] This was a very popular location not only because of the busy bus stop but because there was a store, we called them "candy stores," adjacent to the street corner. There were maybe a hundred or so kids on the corner, and directly in front of me there were two white teenage boys leaning on a fire alarm call box. One of them suddenly pulled the fire alarm and they both fled. About five minutes later, an unmarked police car with two white plainclothes cops pulled up and parked adjacent to the street box.

The cop sitting in the passenger seat made a hand gesture toward the street box, looked directly at me, and said, "Did you pull that alarm, nigger?" As soon as the officer repeated it and began to open his car door, I started to slowly back away into the large crowd. If you attack me with words, I fight you back with words. If you attack me with hands, I fight you back with hands. I quickly blended into the crowd. I saw one of my classmates picking up a rock, so I picked up a rock. We both threw rocks in the direction of the officers and ducked into the corner store. That instigated a storm of rocks and bottles directed at the officers, who, after being struck, retreated into their vehicles and fled the area. The corner quickly cleared out as a large fire truck arrived. The officers never even looked for me, and I didn't care if they did. The incident shaped how I would interpret the concept of armed resistance forever. You see, being called a nigger by a Chicago police officer was not unusual back in 1973, even as a child. Although most of the time it was normally mumbled or mentioned as a side note, this white police officer wanted to let a nine-year old Black male, in his eyes a man-child, know that "nigger" was

Tony Gaskew, *The Ritual of Black Armed Resistance* In: *Wicked Problems*. Edited by: Austin Choi-Fitzpatrick, Douglas Irvin-Erickson, and Ernesto Verdeja, Oxford University Press. © Oxford University Press 2022.
DOI: 10.1093/oso/9780197632819.003.0002

going to be used as a weapon of intimidation and violence on the CPD use-of-force continuum, just like a canine, pepper spray, taser, or AR-15 semi-automatic assault-style rifle. When it happened, I wasn't surprised. I was already culturally resistant to white police officers and their twenty-four-hours-a-day, seven-days-a-week bullshit hustle. In my mind, he was just another chump, in a very long line of chumps with badges and guns talking shit. It didn't scare me. Armed resistance to state oppression was never an ethical dilemma for me; it was a matter of self-preservation.

When I shared this incident with my father, he began laughing and told me something that changed the course of my life forever. My dad explained that his three sons, by our very birthright and spiritual inheritance, or what the Yoruba in West Africa refer to as an *Oriki*, were to be vanguards of justice. He explained that my life's purpose was to engage in a war with white supremacy and that policing in America was the armed wing of white supremacy. As a soldier in the Black Power movement, my dad spoke the language of critical race theory long before Derrick Bell advanced the term throughout the academy.[2] My dad explained that white supremacy must be fought face to face. He explained that I, along with my two brothers, would be entering the battlefield of policing in order to fight white supremacy.

By January 2000, I had spent nearly sixteen years in policing, fighting white supremacy as my father had prophesied. While a significant part of my battlefield experience in policing was in Florida, my brothers had already engaged in nearly two decades of resistance inside the notoriously anti-Black Chicago Police Department. However, once again a conversation with my dad would change the course of my life. My dad asked if I had finished what I started in policing. When I explained that the war against white supremacy never ends, he once again laughed and told me how proud he was of me and my sacrifice for Black liberation. My dad then told me to prepare myself to widen the battlefield in policing. He told me to finish graduate school and to weaponize my intellect, expanding the war against the armed wing of white supremacy from the streets to the classrooms. My dad understood Kwame Ture's vision of Black power and the role education plays in organizing the masses from the *unconscious to the conscious*.[3] I started graduate school at Nova Southeastern University in South Florida, transitioning out of policing in 2004 to complete my doctoral research agenda.

Fast-forward to June 2020 and I'm a tenured Full Professor of Criminal Justice, faculty affiliate in Africana Studies, and Director of the Criminal Justice Program at the University of Pittsburgh, Bradford. In fact, as a Black male, I'm part of a very small fraternity that makes up less than 2% of all full professors within the academic discipline of criminal justice.[4] I had weaponized my intellect, immersing myself in publications, presentations, and revolutionary Black Radical Tradition actions that aggressively challenged every narrative created by the policing culture to criminalize the Black diaspora in America. Although my father has since transitioned into the world of my ancestors, he continues to bring his wisdom to me through the Afrikan ritual of meditation. Thus, I continue to dedicate every moment of my career in the academy fighting white supremacy, following Ture's empathic call by ensuring that every single student, staff member, faculty, or administrator whom I come into contact with is educated not only on the poisonous nuances of white supremacy within American policing but also on the inevitable reality that policing will eventually be dismantled and abolished.[5]

Within this brief essay, using the voices of the Black Radical Tradition, I argue that Black armed resistance is one of the collective rituals of revolution, required to dismantle the institution of American policing. If any ethical dilemma exists involving the use of violence as a universal right of Black self-defense, the burden is firmly placed on the collective backs of white liberals and their choice to either support the state or the Black diaspora in revolution. If history has taught us anything about white liberals and how they prioritize their own self-interests, needs, and power, it is much more likely they will support the former rather than the latter. This theme resonates with my book, *Stop Trying to Fix Policing: Lessons Learned from the Front Lines of Black Liberation*, which describes my organic critical autoethnographic journey across America, asking the Black diaspora one central question: *How do we get policing out of our lives?*[6] This journey unearthed five rituals that can be used for dismantling the institution of policing in America, all ingrained within the Black Radical Tradition of resistance: *speaking the language of police abolition, unfriending policing, decolonizing the state narrative, community self-determination*, and *armed resistance*. The ritual of *Black armed resistance* invokes the universal right of self-defense, which sits at the cosmological center of what Muata Ashby in *Introduction to MA'AT Philosophy* describes as the Kemetic spiritual inheritance of the Black diaspora: *truth, justice, balance, harmony, order, righteousness*, and *reciprocity*.[7] My journey over the past thirty-eight years fighting against the armed wing of white supremacy is a reflection of my spiritual inheritance and my earthy love for the first human beings on this planet.

A History of Black Armed Resistance

The Black Radical Tradition is a 200,000-year-old spiritual inheritance that weaves the ontological journey of Black humanity, Black survival, and Black power. It is the collective consciousness of the Black cultural experience. It is the totality of truth, justice, balance, harmony, order, righteousness, and reciprocity, passed on through rituals, from one generation of Blackness to the next, creating an endless cycle of cosmic resistance to oppression. The Black Radical Tradition has been working to dismantle policing ever since the Barbados Slave Code of 1661 was used to legitimize the first slave patrols that secured the construct of white supremacy in the Carolinas in the late seventeenth century. American policing today is made up of slave patrols. American policing is the armed wing of white supremacy. The Black Radical Tradition has always been at war with policing. Armed resistance is the story of the Black diaspora. Never lose sight of these universal truths. Slave patrols are America's original perpetrators of the homoerotic stop-and-frisk fetish: police carnal violence.[8] Deputized white males, infatuated with holding, touching, and handling captive Black male bodies for consumer fantasization, led to the discovery that there was a price to pay for their savage anti-Black crimes. Kerry Walters,[9] Herbert Aptheker,[10] Richard Price,[11] and Alvin Thompson[12] provided clear evidence that armed resistance to white supremacy shifted the entire narrative of Black liberation. Armed maroon communities would hunt down slave patrols and introduce them to one of the central themes embraced one hundred years later by the Black Liberation Army:

> America must learn that Black people are not the eternal suffers, the universal prisoners, the only ones who can feel pain. Revolutionary violence is, therefore, not a tactic of struggle, but a strategy . . . forcing all those responsible for oppression to realize that they too can bleed, they too can feel our pain. Only when this is realized, will any just and equal decisions be made, will we be conceded our right to self-determination.[13]

To place this narrative into a construct that most white liberals can digest, on October 14, 1964, Martin Luther King Jr. was awarded the Nobel Peace Prize, making him the youngest winner of the prize in its history.[14] In 1959, five years before winning the prize and four years before his "I Have a Dream" speech, MLK articulated the Black Radical Tradition's response to American white supremacy and its evil rhetoric of domination and hegemony, as synergized by Marimba Ani.[15] King's response echoed the voices of Black armed resistance, when he noted one popular view of violence as:

> self-defense, which all societies from the most primitive to the most cultured and civilized, accept as moral and legal. The principle of self-defense, even involving weapons and bloodshed, has never been condemned, even by Gandhi, who sanctioned it for those unable to master pure nonviolence. When the Negro uses force in self-defense, he does not forfeit support—he may even win it, by the courage and self-respect it reflects.[16]

King, who was an avid gun owner, was explicit in his support for armed self-defense.[17] In fact, he even sought a permit to carry a concealed gun in his car.[18] What MLK clearly understood was the exact same thing that maroon communities, Black Civil War veterans, the Deacons for Defense, the Black Panther Party, the Black Liberation Army, and millions of the Black diaspora across America today who own guns clearly understand: the institution of white supremacy fears armed Black people.[19] Always has, and always will. Frederick Douglass's advice that a good revolver was the best response to slave catchers[20] resonates with the Black diaspora today. So does W. E. B. Du Bois's warning that a double-barrel shotgun would be used to spray the guts of a white mob.[21] As does Fannie Lou Hamer's words that she kept a shotgun in every corner of her bedroom and would not hesitate to use them on the first white person to do her harm.[22] Let's not forget Marcus Garvey's voice, that all people have gained their freedom through organized force.[23] Finally, Malcolm X's vision of Black liberation was not against using violence in self-defense, because he believed that armed self-defense is not violence but intelligence.[24] You see, the Black diaspora within the Black Radical Tradition has always affirmed the power of armed resistance and the natural right of self-defense.

The image of Black people with guns sits at the heart of white fear in America. It disturbs the white-centered imagination, whether that's at an academic conference, at an abolitionist meeting, or inside a police department. A few years ago, when I began to share some of my initial research findings during presentations at the Academy of Criminal Justice Sciences and the American Society of Criminology, both considered top conferences within the criminal justice discipline, I was advised by two white faculty peers, who are perceived by other white educators to be prominent scholars in the subfield of abolitionism, to stay away from any topic involving abolitionism and Black armed resistance. Before literally laughing at them, I asked if they would like to amplify their thoughts, and one of them nervously uttered, "What are African Americans going to do with guns? Someone might get hurt." The other paused for a moment and said, "If guns are required, then it's not really abolitionism." Please understand, their views echoed the white liberal agenda I've heard for decades regarding the topic of guns and Black people in America. Well, the Black voices within the diaspora tell a different story. A significant portion of the Black diaspora I've spoken to in this critical autoethnographic journey suggests that without armed self-defense, police abolition would be impossible. Armed self-defense shifts the *policing* narrative of Blackness. Armed self-defense shifts the *public* narrative of Blackness. Armed self-defense shifts the *political* narrative of Blackness. Armed self-defense shifts the *cultural* narrative of Blackness. Armed self-defense shifts the *power dynamics* of Blackness. The harmony of safety, the balance of security, and the reciprocity of self-preservation are the responsibilities of the collective Pan-Afrikan diaspora, not the state.

Armed Resistance Is a Spiritual Inheritance

The foundational core for members of the Black diaspora who support the ritual of armed resistance lies in the historical truth of an Afrikan-centered spirituality. Black people created the concept of justice on this planet. This concept of justice is reflected in the principles of Ma'at. There is no higher human behavioral code that has been found anywhere in history.[25] Ma'at is based on seven core universal virtues: *truth, justice, balance, harmony, order, righteousness*, and *reciprocity*. These virtues are protected by the universal science of self-defense, which consist of three interconnective metaphysical levels: spiritual, mental, and physical.[26]

According to Balogun Abeegunde[27] and T. J. Obi Desch,[28] the Kemites, or ancient Egyptians, were the first people on the planet to apply the science of self-defense, or what is referred to as the Afrikan system of self-defense: *Montu*. The *Montu arts* is what is commonly known today as the martial arts. Montu or Montu-Ra is manifested in Afrikan spirituality as the Kemite symbolic god of war, a warrior who would attack the enemies of Ma'at, the concept of truth that is the cosmic order of the universe. Montu was often depicted as a man with the head of a falcon wearing a headdress of two long plumes, a solar disk, and the double uraeus. According to Baba Ifa Karade,[29] Monique Joiner Siedlak,[30] Jade Asikiwe,[31] and Kaba Hiawatha Kamene,[32] within the Afrikan metaphysical journey of universal self-defense, Ogun is the god of war of the Yoruba people of West Africa. Ogun is the ruler over the elements of nature, specifically steel, with which weapons can be made. Within the Afrikan metaphysical tradition, someone in the position of a police officer would be considered a child of Ogun. For the Pan-Afrikan diaspora, this alone disqualifies American policing from ever being able to sit at the throne of truth and justice.

Today, nearly 20% of the Black diaspora in America own firearms. That's nearly 8.5 million Black firearm owners in America. Another 30% live with someone who owns a firearm. Additionally, nearly 60% of the Black diaspora in America perceive guns as not only a positive thing but, in many cases, a necessity.[33] Members of the Black diaspora in record numbers are now joining gun clubs, going to gun ranges, and participating in competitive shooting events. There are Black-owned gun clubs that hold concealed-weapon registration seminars along with voting-registration campaigns across the country. In fact, in the spirit of Asante Shakur, Black women are now one of the fastest growing demographic groups in America who are purchasing guns.[34]

Black armed self-defense forces an uncomfortable reality for the future of America. This is the reason why America, as a settler colony, has dedicated the past four hundred years trying to disarm Black people.[35] Black disarmament legislation started as early as 1680 in Virginia, which was immediately followed by other states barring gun ownership to Black people, including through the Fugitive Slave Acts, the Black Codes, Jim Crow, the war on drugs, mass incarceration, police militarization, and the creation of legislation prohibiting convicted felons from legally possessing guns.[36] The Sentencing Project estimates that nearly one-third of all Black men in America are prohibited from legally owning a firearm.[37] That's 4 million Black adult men, which does not include the hundreds of thousands of Black women who cannot legally possess a firearm and invoke their natural right of self-defense in America.[38]

According to Christian Davenport, the concept of armed self-defense in Black communities is not new in Black liberation efforts.[39] Advanced by the Republic of New Africa (RNA) as early as 1969, it is gaining renewed interest today. The RNA has long urged the Black diaspora to take up arms as a display of political power, cultural power, and Black resistance. The RNA maintains that a significant number of the Black diaspora in America are already supportive of armed self-defense.

Why Do Guns Matter?

According to Peter Gelderloos, policing weaponizes nonviolence and uses it to exploit the weaknesses of the oppressed masses; nonviolence is a tactic that ultimately results in failure because it is *ineffective, racist, statist, patriarchal, tactically and strategically inferior,* and *delusional*.[40] Gelderloos adds that throughout the history of the United States of America, a blind radicalized adherence to the tactic of nonviolence, or what is rationalized as pathological nonviolence,[41] has created a plethora of leaders, policies, and movements that is simply co-opted by white supremacy.[42]

What we often forget within the realm of the Black Radical Tradition, within the realm of Black radical resistance, is that nonviolence and violence are only tactics, not stand-alone mantras for Black liberation. Tactics, that is, effective tactics, are only actions that produce desired reactions. Effective tactics flow from effective strategies or organized game plans, which flow from effective goals or purposes. You choose the most effective tactics, whether violent or nonviolent, that are strategically organized and planned in order to produce your most desired result.[43] Thus, this leaves us with several critical unanswered questions. If we acknowledge that white supremacy is violence and that policing is the expression of white supremacy's violence, what are the most effective tactics or diversity of tactics, violent or nonviolent, to fight white supremacy and abolish policing? Is armed resistance or the threat of armed resistance, a tactic of violence or nonviolence?

The historical inheritance of the Black diaspora in America is one of armed resistance. We have never critically understood or measured the long-term effect that an armed Black revolutionary mindset has played on the psyche of white America. Can we honestly attribute any success for Black liberation in America to nonviolence? Does anyone really believe that two centuries and hundreds of armed Black revolts[44] and over 200,000 armed Black veterans of the Civil War[45] did not impact every single Black liberation effort, including those framed as reformist or nonviolent, over the past 155 years? Are we to believe that "sit-ins," "hunger strikes," and "petitions" ended the enslavement of 4 million Black people in America? Black people who have historically armed themselves with guns have left us with a formula for liberation. Nearly every so-called nonviolent Black political activist in American history, including MLK, carefully crafted a public narrative that used Black revolutionary armed violence as the rational outcome if their nonviolent reformist demands were not met.

American policing prefers the Pan-Afrikan diaspora to be passive, nonviolent, and unarmed. Nonviolence in the hands of the police has been and continues to be a colonial enterprise. Nonviolence assures a police monopoly on armed violence by creating a worldview where policing is the only legitimate source of armed self-defense. American policing recognizes that organized Black revolutionary activism poses a great threat to its power. The institution of policing has also read Frantz Fanon and understands the science of colonization and social control.[46] Armed violence works. Armed violence is built into the culture of American policing, and thus armed violence is the only language that white supremacy respects or understands.[47]

Is it fair to conclude that Black social movements today, which include Black Lives Matter, are reformist and nonviolent by nature, and that because they are rewarded by white liberals and their media outlets who provide them with funding and choose their leadership to remain nonviolent, they will forever be powerless, forever instruments of American policing? White liberalism is a product of the state, and the state is a product of white liberalism. We must never forget this. White liberalism helps the police legitimize its violence against Black liberation through structural violence. Is the biggest impediment to police abolition today the Black social movements themselves? Gelderloos adds:

> If a movement is not a threat, it cannot change a system based on centralized coercion and violence and if that movement does not realize and exercise power that makes it a threat, it cannot destroy such a system. They [police] cannot be persuaded by appeals to their conscience. We must reclaim histories of resistance, to understand why we have failed in the past and how exactly we achieved the limited success we did. Realizing that nonviolence has never actually produced historical victories towards revolutionary goals opens the door to considering other serious faults of nonviolence.[48]

According to Kuwasi Balagoon,[49] being Black is a political condition that mandates liberation through scientific and organized armed resistance. That is, armed resistance must be a method or theory that involves and evolves into a tactic or practice. Similar to many of the men I spoke with during the research for my book, it was during his time incarcerated that Balagoon began to see clearly just how weak and fragile the nature of policing had become. As an incarcerated student, Balagoon formed political study groups that centered on Black liberation and decolonization efforts. Applying principles of critical race theory, given that white supremacy was a permanent component in American policing, it was the right of the Black diaspora to dismantle, expropriate, abolish, and liberate itself from policing through armed resistance. Balagoon argued:

> It is the right of the people to alter or abolish [policing]. . . . The Slave Patrols were the predecessors of the fugitive squad, Red Squad, and Joint Terrorism Task Forces of today. Black communities must be prepared for an armed struggle and obtain concealed weapons permits. Others will be organized to infiltrate members into the U.S. Armed Forces, correctional departments, security firms, police departments, to obtain hard intelligence, and training as well as access to arms. For the police to stop the movement, they would have to arrest the entire community.[50]

Policing the Police

According to Darrell Miller and Kindaka Sanders, a compelling argument can be made for armed resistance against police violence.[51] Miller and Sanders argue that in *District of Columbia v. Heller*, the U.S. Supreme Court held that self-defense lies at the core of the Second Amendment and that it codifies a natural right, for both individuals and communities, to keep and bear arms for self-defense, which extends to both private and public threats, including self-defense against government violence. Miller and Sanders argue that the Supreme Court—in deeming the Second Amendment a fundamental right in *McDonald v. City of Chicago*—*inadvertently* highlighted its role in protecting newly freed Black people, thereby enabling them to bear arms for defense against rogue law enforcement officers who sought to disarm them. In the *Official Reports of the Supreme Court*, Justice Clarence Thomas wrote about the aftermath of Nat Turner's 1831 slave rebellion in Virginia, "The fear generated by these and other rebellions led Southern legislatures to take particularly vicious aim at the rights of free blacks and slaves to speak or to keep and bear arms for their defense."[52] The Court added that the right to keep and bear arms for self-defense includes the ability to defend oneself against tyrannical local or state violence. Miller explained:

> Among the McDonald Court's reasons was a recognition that, during Reconstruction, local law enforcement was, in fact, behaving tyrannically. Local police and recusant state militias terrorized freedmen, sometimes alone, sometimes in collusion with unofficial citizen patrols and groups like the Klan. If one of the principal aims of the Civil Rights Act of 1866 and the Fourteenth Amendment was to allow freedmen to arm themselves in order to repel unreconstructed Southern law enforcement, then it seems that modern individuals would enjoy a constitutional right to publicly arm themselves in case they need to threaten, to resist, or even to fire upon police officers who violate the law.[53]

While the Supreme Court's interpretation of the Second Amendment establishes that there is a right to use armed resistance against the police, the Court failed to provide any specific guidance, scope, or limitations regarding the contours of such a right, as well as whether this right would extend to third-party defense.[54] Because the legal interpretation is so ambiguous, the overwhelming majority of the states, with strong support from Fraternal Order of Police unions, have criminalized even resisting the police for an unlawful arrest. Today, fewer than thirteen states allow a person to resist an unlawful arrest. However, Miller argues that the U.S. Constitution does provide a refuge: the jury.[55] That is, any claim and evidence of armed self-defense in reaction to unlawful police violence must be presented to a jury, where natural law meets positive law. "The natural right to self-preservation can never be sequestered from the jury."[56]

According to Sanders,[57] due to this legal ambiguity, the Indiana Supreme Court abolished the English common law right to resist an unlawful arrest under *Barnes v. State*. Sanders noted:

> The U.S. Supreme Court has not directly considered the question of whether citizens can use deadly force in defending against unlawful police force. However, in John Bad Elk [*John Bad Elk v. United States*], the Court implied that the defendant could have used deadly force against police officers who were about to use deadly force against him. At common law, a citizen had the right to use deadly force against a police official using

unnecessary, deadly force in effecting a lawful arrest. Most states have not abridged this common law rule.[58]

In 2012, in reaction to what many Indianans perceived as a major infringement on gun owners' right of self-defense, the Indiana State Legislature enacted an unprecedented, first-of-its-kind statute, Indiana Code § 35-41-3-2, authorizing the use of force, with an extension for third-party defense, to include the use of deadly force against public servants acting unlawfully against the persons or property of Indiana citizens. Within this narrative, Sanders argues a compelling issue, specifically as it relates to Black armed resistance: the Supreme Court has opened a Pandora's box of legislation similar to Indiana Code § 35-41-3-2 to be enacted across the nation, where the use or threatened use of defensive force against unlawful police violence may in fact serve as a real deterrent for police misconduct involving excessive force in Black communities.[59]

In doing so, Sanders makes an argument that Black communities across America—as the most targeted population for racialized unlawful police violence in the history of this country, going back to policing's origins as slave patrols and rising through the ranks of the Black Codes, Jim Crow, and the FBI's counterintelligence program (COINTELPRO)—are the most in need of Second Amendment protections and subsequent specific state legislative statutes.[60] New laws are needed to establish the legal right of armed self-defense, including third-party armed self-defense, against tyrannical police misconduct and associated police violence. Additionally, Sanders argues that armed Black community patrols that have a widespread understanding of armed self-defense, like those discussed in this chapter and whose actions are already upheld by the Second Amendment, could be invaluable in curbing unlawful police violence.[61] Sanders notes, "The Second Amendment as recently interpreted in *McDonald* and *Heller* provides the vehicle for the effective use of [armed] community patrols and intervention. State statutes that describe the right to defend in resistance will provide further clarity and thus will help empower citizens to challenge individualized government tyranny in their communities."[62]

Just imagine if similar legislation were passed in every state across the nation. This initiative would have a deep and broad ripple effect on the real-world power relationship between policing and the Pan-Afrikan diaspora. Black armed resistance, supported with groundbreaking state legislation similar to Indiana Code § 35-41-3-2, with a clear understanding of constitutional rights of self-defense against unlawful police violence, could play a significant role in the systemic abolishment of policing in America. Which leaves us with one final lingering challenge. We already have tens of thousands of armed and trained Black police officers in America.

Final Thoughts

If I have successfully argued that Black armed resistance is one of the collective rituals required to dismantle the institution of American policing, then logic demonstrates this is an ethical dilemma for white liberals. The right of self-defense is a universal Karmic virtue. If white liberals want the police to protect their self-interests over the liberation and lives of Black people, then at what cost? Today, the institution of policing, the armed wing of white supremacy in America, finds itself fragile, impotent, and ripe for being dismantled.

The police murders of George Floyd, Breonna Taylor, Ahmaud Arbery, and Rayshard Brooks, all within a few weeks of each other in 2020, have triggered what I describe elsewhere as "a wave of delegitimization."[63] A growing segment of white liberals no longer trusts, respects, or, more important, fears American policing. This is crucial when you understand a key demographic responsible for maintaining the institution of American policing today is white liberals. However, what most people fail to understand is that policing in America runs off an algorithm based on white passive acceptance. That is, policing in America, numbering fewer than 1 million full- and part-time police officers, has no real power unless 200 million white people passively accept and recognize their authority. Any level of active resistance by the white masses overwhelms the ability of policing to maintain any level of control or authority, hence the use of the National Guard and U.S. military as support mechanisms, brought in to save policing during uprisings and demonstrations. When that level of resistance becomes organized and hits critical mass, let's say 5% of the white adult population, policing will be dismantled.

Kwame Nkrumah argued that only through the collective revolutionary lens of the Pan-Afrikan masses can we abolish worldwide systems of oppression.[64] That we must all embrace the struggle of revolutionary armed resistance. That we must know who we are resisting, what we are resisting, when we are resisting, where we are resisting, why we are resisting, and how we are resisting. I hope this essay serves as a starting point for a collective truth, in understanding that the Black diaspora's synergized resistance of American policing is, in fact, a global resistance to all state oppression. That everyone benefits from the truth cemented in the fearless Black Radical Tradition of self-defense. If there is a wicked problem, it belongs to white liberals.

Notes

1. Excerpted and adapted from Tony Gaskew, *Stop Trying to Fix Policing: Lessons Learned from the Front Lines of Black Liberation* (Rowman & Littlefield, 2021). Published by permission of Rowman & Littlefield. All Rights Reserved.
2. Derrick Bell, "A Holiday for Dr. King: The Significance of Symbols in the Black Freedom Struggle," *U.C. Davis Law Review* 17, no. 2 (1984): 433–444; Derrick Bell, *And We Are Not Saved: The Elusive Quest for Racial Justice* (New York: Basic Books, 1987); Derrick Bell, "White Superiority in America: Its Legal Legacy, Its Economic Costs," *Villanova Law Review* 33, no. 5 (1988): 767–779; Derrick Bell, "Racial Reflections: Dialogues in the Direction of Liberation," *UCLA Law Review* 37, no. 6 (1990): 1037–1100; Derrick Bell, *Faces at the Bottom of the Well* (New York: Basic Books, 1992); Derrick Bell, "Racial Realism," *Connecticut Law Review* 24, no. 2 (1992): 363–379.
3. Kwame Ture, *Black Power: The Politics of Liberation* (New York: Vintage Books, 1967).
4. Helen Taylor Green, Shaun Gabbidon, and Sean Wilson, "Included? The Status of African American Scholars in the Discipline of Criminology and Criminal Justice Since 2004," *Journal of Criminal Justice Education* 29, no. 1 (2018): 96–115.
5. Kwame Ture, "Stokely Carmichael 'We Ain't Going,'" speech, (1967). https://www.youtube.com/watch?v=HKP5_qyGs8c
6. Tony Gaskew, *Stop Trying to Fix Policing: Lessons Learned from the Front Lines of Black Liberation* (Lanham, MD: Rowman & Littlefield, 2021).
7. Muata Ashby, *Introduction to MA'AT Philosophy* (Lithonia, GA: Sema Institute, 2005).
8. Tommy Curry, *The Man-Not: Race, Class, Genre, and the Dilemmas of Black Manhood* (Philadelphia, PA: Temple University Press, 2017).
9. Kerry Walters, *American Slave Revolts and Conspiracies* (Santa Barbara, CA: ABC-CLIO, 2015).
10. Herbert Aptheker, *American Negro Slave Revolts* (New York: International Publishing, 1983).
11. Richard Price, *Maroon Societies: Rebel Slave Communities in the Americas* (Baltimore, MD: Johns Hopkins University Press, 1996).
12. Alvin Thompson, *Flight to Freedom: African Runaways and Maroons in the Americas* (Kingston, Jamaica: University Press of West Indies, 2006).
13. Black Liberation Army, *Message to the Black Movement: A Political Statement from the Black Underground* (Chico, CA: Abraham Guillen Press/Arm the Spirit, 1971), 14–15.
14. Martin Luther King Jr., *Why We Can't Wait* (London: Signet, 1968).

15. Marimba Ani, *Yurugu: An African-Centered Critique of European Cultural Thought and Behavior* (Trenton, NJ: Africa World Press, 1994).

16. King, *Why We Can't Wait*, 12–15.

17. Nicholas Johnson, *Negroes and the Gun: The Black Tradition of Arms* (Buffalo, NY: Prometheus Books, 2014).

18. Charles Cobb, *This Nonviolent Stuff'll Get You Killed: How Guns Made the Civil Rights Movement Possible* (Durham, NC: Duke University Press, 2015); Johnson, *Negroes and the Gun*; Akinyele Umoja, *We Will Shoot Back: Armed Resistance in the Mississippi Freedom Movement* (New York: New York University Press, 2014).

19. Robert F. Williams, *Negroes with Guns* (Eastford, CT: Martino, 1962).

20. Johnson, *Negroes and the Gun*.

21. Umoja, *We Will Shoot Back*.

22. Cobb, *This Nonviolent Stuff'll Get You Killed*.

23. Johnson, *Negroes and the Gun*.

24. Williams, *Negroes with Guns*.

25. Asa Hilliard, Larry Williams, and Nia Damali, *The Teachings of Ptahhotep: The Oldest Book in the World* (Scotts Valley, CA: CreateSpace, 1987).

26. Ashby, *Introduction to MA'AT Philosophy*, 18–138.

27. Balogun Abeegunde, *Afrikan Martial Arts: Discovering the Warrior Within* (Atlanta, GA: Boss Up, 2008).

28. T. J. Obi Desch, *Fighting for Honor: The History of African Martial Art in the Atlantic World* (Columbia: University of South Carolina Press, 2008).

29. Baba Ifa Karade, *The Handbook of Yoruba Religious Concepts* (New York: Weiser Books, 1994).

30. Monique Joiner Siedlak, *Seven African Powers: The Orishas* (New York: Oshun, 2016).

31. Jade Asikiwe, *Melanin: The Gift of the Cosmos* (Scotts Valley, CA: CreateSpace, 2018).

32. Kaba Hiawatha Kamene, *Spirituality before Religions: Spirituality Is Unseen Science* (Scotts Valley, CA: CreateSpace, 2019).

33. National African-American Gun Association website, accessed August 22, 2020, https://naaga.co/; Lydia Saad, "What Percentage of Americans Own Guns," Gallup, accessed August 22, 2020, https://news.gallup.com/poll/264932/percentage-americans-own-guns.aspx.

34. National African-American Gun Association website.

35. Quashon Avent, "African Americans and the 2nd Amendment: The Need for Black Armed Self-Defense," *The Carolinian*, February 20, 2019, from https://carolinianu ncg.com/2019/02/20/african-americans-and-the-2nd-amendment-the-need-for-black-armed-self-defense/.

36. Cobb, *This Nonviolent Stuff'll Get You Killed*.

37. The Sentencing Project, "Race and Justice News: One-Third of Black Men Have Felony," October 10, 2017, https://www.sentencingproject.org/news/5593/.

38. Black Demographics, "Black Male Statistics: Population," accessed August 22, 2020, http://blackdemographics.com/population/black-male-statistics/.

39. Christian Davenport, *How Social Movements Die: Repression and Demobilization of the Republic of New Africa* (Boston: Cambridge University Press, 2014).

40. Peter Gelderloos, *How Nonviolence Protects the State* (Boston: Detritus Books, 2018).

41. Ward Churchill, *Pacifism as Pathology: Reflections on the Role of Armed Struggle in North America* (Chico, CA: AK Press. 2007).

42. Churchill, *Pacifism as Pathology*.

43. Black Liberation Army, *Message to the Black Movement*; Jalil Muntaqim, *On the Black Liberation Army* (Montreal: Montreal Anarchist Black Cross, 1997); Jalil Muntaqim, *We Are Our Own Liberators* (Chicago: Arissa, 2010); Kwame Nkrumah, *Handbook of Revolutionary Warfare* (Bedford, UK: Panaf Books, 1968); Thomas Sankara, *Thomas Sankara Speaks: The Burkina Faso Revolution 1983–1987* (Atlanta, GA: Pathfinder, 1988).

44. Kuwasi Balagoon, *A Soldier's Story: Revolutionary Writings by a New African Anarchist* (Montreal: Kersplebedeb, 2019); John Henrik Clarke, *Christopher Columbus and the African Holocaust: Slavery and the Rise of European Capitalism* (New York: A & B Books, 1992).

45. Johnson, *Negroes and the Gun*; Umoja, *We Will Shoot Back*.

46. Frantz Fanon, *The Wretched of the Earth* (London: Penguin Books, 1967).

47. Black Liberation Army, *Message to the Black Movement*; Muntaqim, *On the Black Liberation Army*; Nkrumah, *Handbook of Revolutionary Warfare*; Sankara, *Thomas Sankara Speaks*.

48. Gelderloos, *How Nonviolence Protects the State*, 37–38.

49. Balagoon, *A Soldier's Story*.

50. Balagoon, *A Soldier's Story*, 106–344.

51. Darrell Miller, "Retail Rebellion and the Second Amendment," *Indiana Law Journal* 86, no. 3 (2011): 939–976; Kindaka Sanders, "A Reason to Resist: The Use of Deadly Force in Aiding Victims of Unlawful Police Aggression," *San Diego Law Review* 52, no. 3 (2015): 695–750.

52. Supreme Court of the United States, *Official Reports of the Supreme Court*, vol. 561, part 1 (Washington, D.C.: U.S. Government Publication Office, 2010), 1–476, 585, 1001–1020.

53. Miller, "Retail Rebellion and the Second Amendment," 943.

54. Miller, "Retail Rebellion and the Second Amendment," 943; Sanders, "A Reason to Resist."

55. Miller, "Retail Rebellion and the Second Amendment," 974–976.

56. Miller, "Retail Rebellion and the Second Amendment," 976.

57. Sanders, "A Reason to Resist," 696.

58. Sanders, "A Reason to Resist," 728.

59. Sanders, "A Reason to Resist," 696.

60. Sanders, "A Reason to Resist," 740.

61. Sanders, "A Reason to Resist," 744.

62. Sanders, "A Reason to Resist," 745.

63. Gaskew, *Stop Trying to Fix Policing*, 87.

64. Nkrumah, *Handbook of Revolutionary Warfare*.

2

Challenging the Violence of the State through a Movement to End Poverty

Liz Theoharis and Noam Sandweiss-Back

Social movements in the United States have long confronted that most wicked of dilemmas: how to counter the violence of the state with a genuine vision of human rights, and how to make that vision a viable political project. Today we face these very same questions, and although there are no definitive answers, we see nonviolent collective action as an indispensable tool for solving the misery of our time. To be clear, the kind of nonviolence we envision is not a rarified ethic or passive act. Nor is it in contradiction to self-defense and the long history in this country of armed resistance in the face of state-led violence, as Tony Gaskew vividly describes in chapter 1. Rather, it is the revolutionary nonviolence that has long fueled social movements in this country and is an essential building block of a mass movement of the poor and dispossessed today.

What will it take to build such a movement? As political organizers committed to transformative change for the poor, this is the question that consumes our attention. In chapter 3, Kirssa Cline Ryckman raises compelling points about the likelihood of spontaneous violent collective action among oppressed peoples, even in nonviolent movements, and suggests that there can be short- and even long-term benefits to such acts of violence. There is no doubt that without relief, the pain and anger of immiserated people will be expressed in increasingly confrontational, even violent ways, and these experiences can be instructive for those involved. For example, many of our mentors were first politicized in the flames of uprisings like the Watts Rebellion in 1965. But they will be the first to describe their terror as young people when they discovered that their brief and heroic resistance was outmatched by police and National Guard forces armed with weapons of war. They will also remind us that those uprisings, absent a larger organized movement of poor and dispossessed people, were swept up in a reactionary

Liz Theoharis and Noam Sandweiss-Back, *Challenging the Violence of the State through a Movement to End Poverty*
In: *Wicked Problems*. Edited by: Austin Choi-Fitzpatrick, Douglas Irvin-Erickson, and Ernesto Verdeja,
Oxford University Press. © Oxford University Press 2022. DOI: 10.1093/oso/9780197632819.003.0003

counterattack that by the late 1960s had already begun to undermine the gains made by the Black Freedom Struggle, the Great Society, and the War on Poverty, from which we are just now emerging.

In truth, we are less concerned in this essay about the actions of exploited and impoverished individuals, or even groups of people, than we are about the question of grand strategy for fundamental social change. Ultimately, only a movement that can contest and transform the power of the state will achieve lasting relief for those who suffer under the current economic and political system. This will require a level of mass social awakening and mobilization that, in this country, can be achieved only through a strategy of collective nonviolent action and political organization. Approaching this topic with such a wide frame also means that our treatment of violence must begin not with those who are subjects of the state but with the state itself.

The Greatest Purveyor of Violence in the World

For six weeks in the late spring of 1968, thousands of poor people occupied the Washington Mall in an encampment known as Resurrection City. Two months earlier, Rev. Dr. Martin Luther King Jr. had been murdered on the balcony of the Lorraine Motel while he was in Memphis for a sanitation workers' strike. By then King had been on the road for months building support for the Poor People's Campaign, an escalated project of nonviolent civil disobedience intended to make the national crisis of poverty visible and indisputable. The Poor People's Campaign represented a shift from what King saw as an era of civil rights and reform to a new era: one of human rights and revolution.

On June 19, 1968, two days before police raided and destroyed Resurrection City, tens of thousands of poor people marched to the Lincoln Memorial for the Solidarity Day Rally. It was also Juneteenth, the historic day in 1865 when enslaved people in the far reaches of the defeated Confederacy learned that they had been emancipated. The weight of distant and near history hung heavy but hopeful as speakers considered the country as it was and as it could be. Coretta Scott King spoke with the particular clarity of a widow unwilling to fall into the trap of a personal debate on the violence done to her family, entirely committed as she was to uprooting the evils of racism, poverty, and militarism. She reminded the nation that it was host to deeper, more devastating forms of violence:

In this society, violence against poor people and minority groups is routine. I remind you that starving a child is violence. Suppressing a culture is violence. Neglecting school children is violence. Punishing a mother and her child is violence. Discrimination against a working man is violence. Ghetto housing is violence. Ignoring medical needs is violence. Contempt of poverty is violence. Even the lack of willpower to help humanity is a sick and sinister form of violence.[1]

The words of Coretta Scott King were a significant corrective to the national discourse of her day. For years, leaders within the Black freedom movement—even in their disciplined nonviolence—had been painted in distorted, violent, and hyperracialized terms as instigators, outside troublemakers, menaces, and far worse. Uprisings in cities across the country had erupted in the summer of 1967, continuing into 1968, and were met with heavily armed and hardline police forces and messages like that of the Miami police chief Walter E. Headley, who infamously told reporters, "When the looting starts, the shooting starts."[2]

Amid this social upheaval, Coretta Scott King insisted on a different starting place for understanding violence and the mass suffering that swirled around her. In just a few sentences she clarified an entire political and strategic worldview, one that her husband had named in his "Beyond Vietnam" sermon a year to the day of his assassination: the state and its apparatuses were the purveyors of the most extraordinary forms of violence, sometimes directly but more often through regressive policies that separated people from the basic resources necessary to sustain life, like housing, healthcare, wages, and education. She explained that for every bullet and baton there was a piece of legislation and a government program that normalized poverty and proliferated death. This, she asserted, was what violence looked like at a societywide level. Its opposite was what she and countless others had long been fighting for: a nation that could feed, house, educate, and provide for every person.

Today when we are asked to clarify our position on nonviolence, we insist on the same starting place that Coretta Scott King chose. As organizers committed to building a movement to end poverty, our strategy begins with how we understand the structures of violence in the United States and what it will take to restructure society to value and defend all life.

As we write this essay, the United States is in the midst of the COVID-19 pandemic, which has been an accelerant of the death-dealing systems that have long haunted our country. Before the pandemic, there were 140 million people who were poor or a $400 emergency away from poverty, and a study from the Mailman School of Public Health at Columbia University found that seven hundred people died every day from poverty conditions.[3] For decades, a neoliberal campaign led by the rich—decreased public spending, privatization, deregulation of banking and financial markets, attacks on organized labor, slashed taxes, and stagnated wages—produced a few billionaires while swelling the ranks of the poor, disproportionally people of color but made up of wide swaths of every demographic. Poverty, hunger, sickness, and homelessness were rampant and growing before the pandemic.

Now we are staring down a prolonged economic crisis in which many millions more are being left with little or nothing. The solutions on offer betray a prevailing belief that people are poor because of moral or personal failures and that poverty can at best be managed but never abolished. Implied by and crucial to this belief is a dogmatic faith that the wealthy are the key to national prosperity. It is telling that as we enter a new decade that seems to promise nothing but rising inequality and temperatures, both political parties remain trapped by the political and economic frameworks of fifty years ago. Democrats and Republicans differ here and there, but leaders across the aisle essentially adhere to trickle-down economics and robust welfare programs for the rich. In a 2020 report, Phillip Alston, former United Nations Special Rapporteur on Extreme Poverty and Human Rights who led multiple fact-finding missions to the United States, explained the consequences well: "The overwhelming success of the ideological campaign that supports neoliberal policies is that it has succeeded in convincing people that those in poverty have no one to blame but themselves, while supporting the notion that trickle-down policies will address it."[4]

America's recent history is bloodied with examples. In the wake of the Great Recession, the very speculators and financial institutions that were responsible for the devastating crash were bailed out (in this case by the Obama administration). Twelve years later, in March 2020, the Federal Reserve materialized over a trillion dollars to buoy Wall Street in the days immediately following the COVID-19 shutdown. In April 2020, a bipartisan relief bill appropriated billions of dollars more in benefits to the largest corporations, while offering threadbare and temporary support to most citizens and completely locking out millions in undocumented communities, homeless encampments, prisons, and more.[5] The lethal consequences of a shredded social safety net only sharpened as the months unfolded, deaths spiked, and unemployment ballooned.

These unlivable conditions and the government positions that created them are violence of the highest order. In the richest country in the history of the world, we are living in an era of abandonment amid abundance. But just because poverty and interlocking ills like systemic racism, militarism, and climate change have become endemic to our lives does not make them inevitable or permanent. These systems of violence are based in policy decisions that can be undone and replaced by a shift in national priorities.

That kind of shift has only ever come through mass social awakening, emerging out of movements that are able to build irrefutable power and rouse the conscience of the nation. Our assessment is that such a movement today depends on the ability of poor and dispossessed people to come together across the lines that have divided them, and to bring larger sections of society to a more just vision for the nation.

If this is the goal, nonviolent action and civil disobedience are the political strategies needed to win. Our adversaries are better funded, trained, organized, and armed. They wield the combined power of Wall Street, media, the academy, police, military, and government. A ground-up movement will never succeed if it tries to match the rich and powerful on their terrain and on their terms. But we hold one key advantage: those who suffer from the status quo, who are considered to live in the margins, actually make up the vast majority in this country. The poor are a sleeping giant whose strength lies in their numbers and ability to unite and rally all people of conscience into moral action.

This country's history is punctuated by social movements that have been willing to challenge unjust laws and use the alchemy of nonviolent action to transform hurting and struggling people into agents of profound change. Different moments have called for different forms, from labor strikes in factories and fields to groundbreaking occupations of public land, boycotts, sit-ins, housing takeovers, and major mobilizations. Taken together, this is the kind of strategic nonviolent civil disobedience that Dr. King spoke about a few months before his assassination:

There is nothing wrong with a traffic law which says you have to stop for a red light. But when a fire is raging, the fire truck goes right through that red light, and normal traffic had better get out of its way. Or, when a man is bleeding to death, the ambulance goes through those red lights at top speed.

There is a fire raging now for the Negroes and the poor of this society. They are living in tragic conditions because of the terrible economic injustices that keep them locked in as an "underclass," as the sociologists are now calling it. Disinherited people all over the world are bleeding to death from deep social and economic wounds. They need brigades of ambulance drivers who will have to ignore the red lights of the present system until the emergency is solved. Massive civil disobedience is a strategy for social change which is at least as forceful as an ambulance with its siren on full.[6]

Brigades of Ambulance Drivers

In 2018, fifty years after the rise and fall of Resurrection City and the death of Dr. King, we helped to launch an organizing project of the poor through the largest coordinated wave of nonviolent civil disobedience in twenty-first-century America. Over six weeks between Mother's Day and the Summer Solstice, people in thirty-five states flooded state capitals and Washington, D.C. We took over legislative buildings, government offices, and public space, sometimes in cities that had never experienced such activity. The aim was to break through the moral narrative of the nation by dramatizing the injustices of police violence and the great contradiction of poverty amid plenty, to lay the early groundwork for a national movement that could galvanize people into long-term action.

Those first weeks were a leap of faith and a radical experiment, and although our efforts were often tenuous and uneven, we succeeded beyond our expectations. Thousands volunteered for nonviolent civil disobedience and were arrested and jailed; the two of us were arrested on the steps of the U.S. Supreme Court protesting voter suppression and its connection to poverty, labor rights violations, and inequality. Across the country, people felt the midnight stirrings of a sleeping giant. Tens of thousands showed up in support, many who had spent their whole lives organizing and many more who were being politicized for the first time. Together, we had all founded the Poor People's Campaign: A National Call for Moral Revival.

From its inception, this campaign has been grounded in what we call "fusion politics." We understand that the poor have long been not only unorganized but purposefully disorganized through divide-and-conquer strategies and structures of violence and dehumanization like white supremacy, systemic racism, and other forms of prejudice and hate. We also believe that poor people of all backgrounds can come together to organize on the basis of shared conditions and interests—because it's happened throughout history and we've seen it in our work today. In fact, we hold that the unity of the poor is the path to a genuine era of human rights and revolution, as Dr. King once imagined. The idea of fusion politics itself echoes earlier chapters of political reckoning and transformation in this country. From Reconstruction after the Civil War through the 1880s and 1890s, newly emancipated Black people built powerful, if fragile, alliances with poor whites to seize governing power. Fusion parties expanded voting rights, access to public education, labor protections, fair taxation, and more, all across the South. In North Carolina, they went so far as to rewrite the state constitution in 1868, codifying the right of all citizens to "life, liberty, and the enjoyment of the fruits of their own labor."[7]

The new Poor People's Campaign also developed out of the fusion of different organizing histories that had employed mass collective action and nonviolent civil disobedience to advance the fight for justice. These movement streams, some well-known and others less so, all helped to alter the political landscape of their time and are rich with strategic lessons for today.

One stream wound its way through the South and the Forward Together Moral Movement in North Carolina, where an unlikely coalition of tens of thousands won major legislative and legal victories and unseated a governor who denigrated the poor and people of color. What began in 2013 as a demonstration by seventeen people outside the statehouse in Raleigh quickly escalated into a weekly protest movement that captured the imagination of the public and brought thousands together in nonviolent direct action. They advocated a breathtaking moral agenda and developed creative and explosive forms of nonviolent civil disobedience that rippled from the Piedmont to Appalachia and beyond. Under the direction of leaders like Rev. Dr. William J. Barber II, the Moral Movement opened the minds of millions to the hope of revolutionary change and the possibility that such a change could once again come out of the South.

Another current flowed from the legacy of welfare rights and homeless organizing in the 1980s and 1990s, which traced its lineage back to multiracial industrial organizing in the mid-twentieth century and to the Great Depression. Organizations like the National Welfare Rights Union and the National Union of the Homeless grew and spread in response to the neoliberal offensive of the Reagan administration and decades of attacks on the poor. At the time, federal, state, and municipal welfare systems were being stripped of funds, as was the public housing system. As housing stock diminished, the evisceration of the social safety net made thousands homeless. The patchwork and paternalistic shelter system was permanently overwhelmed, and families were violently torn apart by government agencies like the Department of Children and Family Services. Families doubled and tripled up, lived in cars and alleyways, and sometimes moved into empty houses and onto vacant land. This was an organic and spontaneous form of nonviolent civil disobedience in a nation where it is illegal to occupy abandoned buildings but legal to die on the streets.

Within these miserable conditions and acts of nonviolent individual perseverance, some saw the potential to move from simply surviving to building power. Out of the shelters and from the streets, poor and homeless people began to organize themselves into projects of mutual aid and solidarity, forming local organizations and homeless unions. By the late 1980s, the National Union of the Homeless was formalized, with close to thirty thousand members in twenty-five cities.[8] With a powerful and growing base, organizers across the country escalated their efforts, including waves of coordinated and nonviolent takeovers of vacant, federally owned buildings. These takeovers met the material needs of some at a time when the government had abdicated its responsibility to protect and provide for its poorest citizens. They also served as a campaign of mass popular education, naming and demonstrating the violence of homelessness in cities where there were more unoccupied publicly owned houses than there were homeless people.

Homeless Union leaders were also students of history. Through study and dialogue with elders in the movement, they understood that their housing takeovers were part of a long tradition of nonviolent survival struggles. They knew about the Hoovervilles of the Great Depression, shanty towns that had cropped up in cities across the country and that didn't look unlike their own tent cities. They also studied the Unemployed Councils of the early 1930s, which fought for relief for unemployed workers and prevented or reversed thousands of evictions and gas and electricity shutoffs. These multiracial councils, organized by the Communist Party, held many lessons on how to develop leaders, build mass organizations of the poor, and take strategic nonviolent direct action, even, or especially, when it meant breaking the law.

The 1932 Ford Hunger March in Detroit was a prime example. Unemployed Councils across the city, which had largely been separated by ethnic and racial group, coalesced in early March of that year to take the fight around hunger to the Ford River Rouge Plant, which had laid off the majority of its workforce. On the border of Detroit and neighboring Dearborn, over ten thousand marchers were met by armed police and Ford's private security force. When the unemployed workers refused to stop, they were sprayed with machine guns, killing four and injuring dozens. It led to the largest public funeral in the history of Detroit at the time, attended by members of the Unemployed Councils and residents of the city who could no longer look away, and sowed the seeds for labor organizing in Detroit's industrial plants for decades to come.[9]

Later that same year, a different project of the poor broke out. In late July 1932, over forty thousand unemployed veterans of World War I arrived in Washington, D.C., to demand the early redemption of bonuses that had been promised to them but were redeemable only after 1945. By then, the economic crisis was spiraling but the government refused to take any meaningful action to alleviate the suffering of millions. The Bonus Army, as the veterans called it, collected many of the fraying threads of the American tapestry: Black and white veterans of the nation's largest imperial exploit, wives and partners, children, and more. They made camp on seized public land just across the water from federal offices and held regular nonviolent marches and rallies. During the encampment, the House of Representatives passed a bill to deliver the bonuses, only for the Senate to vote it down. Eventually, President Herbert Hoover ordered the U.S. Army to tear down the camp. The violent treatment of poor and war-weary veterans was a lightning rod for the public, and Hoover would lose the presidency to Franklin D. Roosevelt later that year, setting the stage for a decade defined by militant organizing and major shifts in national policy.[10]

These often-untold histories of nonviolent struggle inspired the Homeless Union in the 1980s and 1990s, which would go on to secure guarantees from the federal government for subsidized housing and protections for the right of the homeless to vote. Despite these visionary efforts, the Homeless Union went into decline by the mid-1990s, undermined by the "chemical warfare" of the crack epidemic, the march of gentrification, and the successful co-optation of some of its leaders. Organizers emerged from the Homeless Union with a hard-won lesson: it had been a strategic mistake to isolate the struggle of the homeless from the struggles of the larger swaths of the poor. In a time of widening poverty, it had become clear that the position of the homeless could not be separated from other poor communities. Nearly three decades later, there could be no greater confirmation of this reality than the arrival of the COVID-19 pandemic and the ensuing economic crisis, which ripped across the country like wildfire.

You Only Get What You're Organized to Take

In April 2020, Senate Majority Leader Mitch McConnell told the press what would happen to proposals around environmental protections and universal healthcare if the Republicans won in the next election: "If I'm still the majority leader in the Senate think of me as the Grim Reaper. None of that stuff is going to pass."[11] In the growing shadow of the worst public health crisis in at least a century, McConnell mocked the prospect of a guaranteed right to healthcare and a livable planet by comparing himself to the harbinger of death. McConnell committed himself and his party to perpetuating the violence that Coretta Scott King spoke of decades earlier.

Two months later, the Poor People's Campaign, adjusting to the need for physical distancing, held the largest digital assembly of poor people in American history, bringing 3 million online and broadcasting via phone and radio to homeless encampments and other forgotten corners of the country. During the assembly, we heard from organizers on the front lines: residents in California leading rent strikes and homeless people again taking over abandoned homes, low-wage workers across the South engaging in labor strikes, leaders in Vermont and Pennsylvania fighting for healthcare with a Nonviolent Medicaid Army of the Poor, and more. We also released a new policy platform that we believe, if enacted, can heal and transform the nation. In its preamble, the Campaign lists a series of orienting principles, which begins, "We need a moral revolution of values to repair the breach in our society."

In this essay, we have discussed nonviolent action and civil disobedience from the vantage point of history and strategy, but we hope it is clear that we offer this argument in service of a larger moral vision for the country and the world. Debates about violence and nonviolence become abstract and fruitless if they are not dedicated to a future in which violence no longer exists and people have access to the resources they need to live fully and abundantly. If we are serious about winning, we must invest all our energy into forcing this nation to undergo a moral revolution of values, one that recenters the attention of society around the needs of the poor and all who are oppressed. With human and environmental catastrophe encircling us, anything less than that forfeits the health of entire generations not just in the present but far into the future.

Nonviolent collective action is our clearest pathway to moral revolution in this country. We are reminded of Dr. King's Nobel Peace Prize acceptance speech in Oslo, three years before his assassination:

> Negroes of the United States, following the people of India, have dem-onstrated that nonviolence is not sterile passivity, but a powerful moral force which makes for social transformation. Sooner or later all the people of the world will have to discover a way to live together in peace, and thereby transform this pending cosmic elegy into a creative psalm of brotherhood.[12]

Frederick Douglass was fond of saying that those who feel the burden of injustice will always be the first to strike out against it, out of a place of life-saving necessity. Today it is also true that the poor have the least to lose from big changes and the most to gain. This suggests that those who are excluded, exploited, laid off, and locked up are best placed to be a leading social force, to spur not only their own communities into action but the millions more who also gain nothing from the current system and are given just enough to maintain the false promise of stability.

To win such a battle will require a mass movement that can imagine a fun-damentally different moral narrative about not just what is necessary but what is possible, right now. In the face of deep suffering, the wealthy and ruling elite have always tried to brand the grim reaper's scythe as a tool of pragmatism and moderation. When that has failed, they have used physical force to bring unruly communities to heel. Both methods mean only vio-lence and death.

On the other hand, organized communities among the poor and dispos-sessed have surfaced and fought for solutions that uphold the sanctity of life. And throughout history, as the poor have realized the power of their num-bers, strategically utilized nonviolent action and civil disobedience, and succeeded in rallying society around a moral call, they have won. In the Poor People's Campaign, we like to say that "everybody's got a right to live." As we endeavor to make this a reality, elders in the movement often remind us of an older slogan from the National Union of the Homeless: "You only get what you're organized to take."

Notes

1. Manseen Logan, "3 Speeches from Coretta Scott King That Commemorate MLK and Cement Their Family's Lasting Legacy," *Blavity*, January 23, 2020, https://blavity.com/blavity-original/3-speeches-from-coretta-scott-king-that-commemorate-mlk-and-cement-their-familys-lasting-legacy?category1=culture.
2. Michael Wines, "Looting Comment from Trump Dates Back to Racial Unrest of the 1960s," *New York Times*, May 29, 2020, https://www.nytimes.com/2020/05/29/us/looting-starts-shooting-starts.html.
3. Shailly Gupta Barnes, Aaron Noffke, Sarah Anderson, Marc Bayard, Phyllis Bennis, John Cavanagh, Karen Dolan, Lindsay Koshgarian, Sam Pizzigati, Basav Sen, and Ebony Slaughter-Johnson, *The Souls of Poor Folk: Auditing America 50 Years after the Poor People's Campaign Challenged Racism, Poverty, the War Economy/Militarism and Our National Morality* (Institute for Policy Studies, 2018); Sandra Galera et al., "Estimated Deaths Attributable to Social Factors in the United," *American Journal of Public Health* 101, no. 8 (2011): 1456–1465.
4. Phillip Alston, *The Parlous State of Poverty Eradication,* report to the United Nations Human Rights Council (New York: United Nations, July 2, 2020).
5. Poor People's Campaign, "Who Is Left Out of the COVID-19 Legislation," April 2020, https://www.poorpeoplescampaign.org/wp-content/uploads/2020/04/Who-is-left-out.pdf.
6. Martin Luther King, *Trumpet of Conscience* (New York: Harper and Row, 1967).
7. William J. Barber II, *The Third Reconstruction: Moral Mondays, Fusion Politics, and the Rise of a New Justice Movement* (Boston: Beacon Press, 2016).
8. National Union of the Homeless, *National Organizing Drive of the National Union of the Homeless,* pamphlet, personal archive of Willie Baptist, 1988.
9. General Baker, "History of Struggle in Detroit," lecture organized by the Poverty Initiative Poverty Scholars Program at the United States Social Forum, Detroit, MI, June 24, 2010.
10. Mickey Z, "History of the Bonus Expeditionary Force (BEF) or Bonus Army," Zinn Education Project, accessed November 22, 2020, https://www.zinnedproject.org/materials/bonus-army.
11. Camilo Montoya-Galvez, "Think of Me as the Grim Reaper: McConnell Vows to Thwart Democratic Proposals," CBSNews, April 22, 2019, https://www.cbsnews.com/news/mitch-mcconnell-vows-to-be-the-grim-reaper-to-thwart-all-democratic-proposals/.
12. Martin Luther King Jr., "Acceptance Address for the Nobel Peace Prize," Oslo, December 10, 1964, Martin Luther King, Jr. Research and Education Institute, Stanford University,. https://kinginstitute.stanford.edu/

3

Is Violence the Answer?

A Pragmatic Approach

Kirssa Cline Ryckman

On a Saturday in late May, I woke up to news of spontaneous collective violence in downtown Tucson, Arizona, and images of protesters being pelted with pepper bullets, broken windows, and Dumpsters set ablaze. Four days earlier, a Minneapolis police officer, Derek Chauvin, killed George Floyd, sparking an extraordinary wave of protests across the United States that, while predominantly nonviolent, was punctuated by acts of violence and vandalism. Here in Tucson, our local Black Lives Matter chapter made the explicit decision to hold virtual events to promote Black healing rather than organize in-person protests. I was humbled and encouraged by the approach the organizers took—they fully recognized the pain and deep grievances felt by the Black community but sought to promote the movement's goals while protecting individual members and participants in a charged environment made doubly difficult by the ongoing COVID-19 pandemic.

The news of violence cut deeply. It undermined the purposeful and deliberate efforts of our local chapter and painted our community's activism as another example of senseless violence. Restaurants and shops that had just reopened after weeks of pandemic-mandated lockdown became collateral damage. The need for cash-strapped businesses to fix and renovate further threatened the well-being of employees already in a state of great uncertainty and turmoil. I was grateful that we suffered only graffiti and broken glass rather than serious injuries and deaths, but even so, I found myself decrying the vandalism and calling for nonviolent approaches. I wasn't alone.

Over time I began to question this impulse. Collective violence had negative consequences in my community and across the United States, but it also got the events in the news, encouraged discussion, and started campaigns for actual change. I became increasingly unsure about my conviction surrounding the primacy of a nonviolent approach.

Kirssa Cline Ryckman, *Is Violence the Answer?* In: *Wicked Problems.* Edited by: Austin Choi-Fitzpatrick, Douglas Irvin-Erickson, and Ernesto Verdeja, Oxford University Press. © Oxford University Press 2022.
DOI: 10.1093/oso/9780197632819.003.0004

I came to appreciate that my uncertainty was the result of a moral dilemma. Avoiding violence can put activists at risk of repression and injury, as we saw time and again during the Black Lives Matter protests. Limiting a movement's tactical range might continue the suffering of people whose grievances are at the heart of the movement by forgoing the potential benefits of such inflammatory acts. Given these consequences, I had to ask whether collective violence can be justifiable, whether violence might be the answer.

The Moral Dilemma of Nonviolent Discipline

Activists have long emphasized the morality of nonviolent methods. The Gandhian tradition of *satyagraha* seeks to achieve change without harming your opponent, to change their heart and mind through compassion and patience rather than coercion. Martin Luther King Jr. consistently emphasized the importance of using moral means in the fight for just goals, viewing nonviolent resistance as the morally superior approach specifically because it does not physically harm your opponent.

This tradition recognizes that nonviolent action requires self-sacrifice and entails risk due to the need to be patient in the pursuit of changes, as well as the serious danger that nonviolent discipline generates for those who stand up to injustice that is backed by coercive force. Nonviolent discipline requires that dissenters remain calm, resist the impulse to run, and, above all, never use violence. Beyond the short-term threat of repression, in the long term limiting the movement's tactical repertoire to nonviolent tactics can concede defeat, stifle progress, and perpetuate grievances. Gandhi himself believed nonviolent resistance was the best method, but he also thought that violence was better than doing nothing.

To be sure, the adoption of violent tactics invites greater costs. There is a significant likelihood of being killed in an armed battle against the state, and research consistently shows that violent movements generate higher levels of repression than their nonviolent counterparts.[1] Yet combatants go into a situation armed, ready to face violence, and trained to stand up to coercive force. Protesters and demonstrators do not. This begs the question of whether nonviolent resistance is worth the short-term costs of repression, given that protesters have no weapons to stand up to state violence. Furthermore, is it worth the potential long-term costs of failure, and what is given up by eschewing violence?

The Long Term: Achieving Movement Goals

To unpack this dilemma, I begin with a consideration of why nonviolent action is presented as both the morally and strategically superior approach to achieve the movement's long-term goals. King described himself as being both "morally and practically" committed to nonviolent resistance,[2] and important to his conviction is the recognition that civil disobedience is not a "do nothing" strategy; it requires resistance, action, and noncooperation with systems that are viewed as evil or unjust. It challenges systems of oppression because the cooperation and compliance of ordinary civilians are necessary for the functioning of the state. Nonviolent movements also have what Erica Chenoweth and Maria Stephan term a "participation advantage."[3] The barriers to participating in a nonviolent protest movement are considerably lower than in a violent campaign, and the larger a nonviolent movement, the more costs can be forced on the state to earn meaningful accommodations or outright capitulation.

Nonviolent action is also valuable because it is hard to successfully coerce. When the state uses violence against nonviolent protesters, it can appear unjustified and a gross overreaction, effectively generating support for the nonviolent actors and eroding support for the state. Activists discuss the beneficial optics generated by nonviolent discipline: when demonstrations remain peaceful, the stark contrast between the violence of the state and nonviolent action of the protesters undermines government legitimacy and encourages backlash. Regimes likewise rely on coercive forces, who may stand down and refuse to repress if they are persuaded by protesters or otherwise unwilling to use violence on peaceful demonstrators. Nonviolent tactics thus pose a critical problem since the state's primary tool can backfire and become a liability.

Conversely, when protesters use violence the government can claim that they are up against criminals, terrorists, or rogues, justifying the use of repressive measures. At the same time, the presence of violent elements tends to reduce the participation advantage that nonviolent movements enjoy, especially among moderates and weak supporters.[4]

There is also evidence that the use of violence can brand a movement in the long term and have lasting negative impacts. The First Intifada in Palestine was a nonviolent movement, yet the images most famously associated with the campaign show Palestinians throwing rocks at Israeli troops and armored vehicles. In their case study of the First Intifada, Chenoweth and

Stephan devote their introduction to appeasing skeptics and showing that this was, in fact, a nonviolent movement.[5] When a movement is so widely, even if incorrectly viewed as violent, it can lose international and domestic support that is critical to its success. In another example, the violent, race-related demonstrations of the 1960s and 1970s in the United States appear to have reduced income and property values for Black Americans for decades to come;[6] increased police killings in general, but especially of Black individuals;[7] and generated a white backlash against civil rights progress.[8] As King asserted, "Every time a riot develops, it helps George Wallace [the segregationist presidential candidate]."[9] Following the period of collective violence, not only did the civil rights movement stop achieving concessions from the government in the short term; it faced serious long-term setbacks that necessitated today's Black Lives Matters movement.

Are There Benefits of Violence?

There is good evidence that nonviolent tactics generate benefits for collective action campaigns, while violence can create significant costs. The other side of this question is whether the use of violence can generate benefits that outweigh those costs, leading to success for a protest movement. Before answering this question, it will be useful to clarify what "violence" is in this context. In particular, I am interested in the use of "collective violence," which consists of actions that are carried out by one group against another group, are intended to cause property damage or physical injury, and are in service of the nonviolent campaign's political goals. This approach omits acts of opportunistic violence and individual crime.

Collective violence can be further differentiated by examining the group behind the event. Spontaneous collective violence describes events that arise without a clear leader, while planned collective violence describes events that are planned by a group that is *not* primarily oriented toward the use of violence. These forms of collective violence are then carried out by demonstrators or nonmilitants, individuals who are not members of an armed organization. This type of collective violence is often labeled "rioting" by popular sources, but the term "riot" is heavy with normative connotations, and its popular use often stands in contrast to both legal and scholarly definitions.[10] As a result, I use the terms "spontaneous" and "planned collective violence" to differentiate these actions from acts of crime and events led by organized militant groups.[11] Second, violence can also be carried out by radical flanks. These are primarily violent groups, such as a militia or militant wing, that have the same overarching goals as the larger nonviolent movement.[12] To answer the question of whether violence can generate benefits to a nonviolent movement, I consider demonstrator collective violence and violence carried out by radical flanks in turn.

Demonstrator Collective Violence

There is commonly cited evidence that collective violence carried out by demonstrators encourages positive change. In the United States, the violent demonstrations of the 1960s led to the creation of the Kerner Commission and its recommendations for local police departments, such as the hiring of minority officers and residency requirements for officers. The 1992 Los Angeles Riots are also frequently cited for their role in reshaping the LAPD, refocusing the department on community policing. Reforms like these seem to be based, at least in part, on the state's recognition that violence could continue or even intensify if left unanswered. This logic underpins research on the connections between collective violence and democratization, and scholars have found that unarmed collective violence, an analytical category close to demonstrator collective violence, is positively associated with democratic liberalization.[13] This link between violence and positive, democratic outcomes, however, is not definitive. Most important, it is not clear if democratization is driven by the violence itself, or if violent acts in conjunction with mass mobilization are spurring change. In other words, it is not clear if movements succeed because of collective violence or despite it.

Although there is some evidence that demonstrator collective violence can encourage positive change, it is not definitive, and there are many negatives to contend with. The immediate costs of damaged property and lost lives, loss of support for the movement, and long-term economic and political impacts make this hard to recommend as a purposeful strategy, and indeed, *planned* collective action is relatively uncommon. According to the Social Conflict Analysis Database, which records collective action events like protests and strikes over a nearly thirty-year period across Latin America and Africa, less than 10% of events coded as "riots" were planned. The rest arose spontaneously. When asked about the value of riots, Sherry Hamby put it well: "They work sometimes, but aren't great solutions."[14]

Radical Flanks

Radical flanks are distinguished by their organization and intent: a radical flank is a group that has the same goals of the nonviolent campaign but is organized for violence as a militia or military wing. The impact of radical flanks is contentious. While there are many reasons to expect that radical flank violence will have a negative impact on a social movement, examples abound of positive radical flank effects. From the U.S. civil rights movement to Euromaidan in Ukraine and the anti-apartheid movement in South Africa, violence appears to play an important role in advancing short-term aims, like gaining acceptance of nonviolent tactics and increasing domestic or international support, as well as the long-term political objectives of the campaign.

There are several explanations for these positive radical flank effects. A radical flank can make the nonviolent faction of a movement appear less threatening and more reasonable, which can allow moderates to present themselves as the side that deserves support and material resources from both third parties and their own constituents. The threat of violence can also generate a crisis for authorities, elevating the nonviolent faction to the negotiation table as the only viable partner to stave off large-scale rebellion or civil war.

While case study evidence suggests that radical flanks can work to generate positive effects, the results of quantitative studies are mixed. On average, radical flanks do not seem to have a positive or negative impact on movement success, but the use of violence depresses the level of participation.[15] Since participation is strongly linked with success, on balance radical flanks are seen as more of a liability than an asset. Limited terrorism, carried out by groups with the same goal as the larger movement, can also encourage benefits *if* the movement is centralized and has strong leadership.[16] In these cases, leaders can keep their members from shifting to violence, allowing them to use the threat of violence to win concessions. Taken together, then, radical flanks are either a net negative or only conditionally positive.

The conclusion from Herbert Haines is illustrative.[17] He argues that positive radical flank effects are very hard to bring about intentionally, because it requires cooperation between groups that fundamentally disagree on tactics, the strategic move could backfire if exposed, and it requires the radical faction to fall in line behind the nonviolent campaign and sacrifice its own objective. He ends with the following: "It is reasonable to assume that positive [radical flank effects] are almost always *unintended* results of movement fractionalization."[18]

The Short Term: The Reality on the Ground

In the long term, nonviolent methods appear to be the better strategy overall, in terms of both morality but also effectiveness, but this comes with the risk of harm to activists in the short term. Is it justified to use violence for self-defense? This is discussed extensively by Tony Gaskew in this volume, who argues for self-defense in the form of an armed resistance for Black Americans in the United States. This is the argument advanced by Kai Thaler, who rightly points out that it is unfair to ask more of unarmed protesters than we do of our security forces.[19] Demanding nonviolence from protesters but applauding police for using "nonlethal" force sets up a double standard that excuses state violence while punishing protesters seeking self-defense. Importantly, this can be a question about the overall strategy to obtain movement success (for example, whether the time has come for an armed resistance), but here I am focused on the question of whether individual-level preservation can take precedence over collective goals.

One path forward is to focus on training in order to help protect protesters. Activist training often covers information on how to stay safe in the face of repression, including learning specific strategies like putting packs on your back and sitting in circles to avoid police dogs or horses, how to isolate hecklers and provocateurs, or how to use medical kits.[20] Such training can give protesters greater confidence and lessen the fear of nonviolent tactics, knowing that they have agency even in the face of violence. Looking across the training used by a variety of movements, Nadine Bloch further notes that a common result of training is that "the safety of the protesters was enhanced because the authorities ultimately trusted their adherence to nonviolent discipline."[21] This suggests that nonviolent resistance carries an important signaling role, allowing security forces to keep their cool knowing that protesters will remain peaceful.

Training on nonviolent discipline can mitigate the risks of repression, but training can do only so much against extreme coercion. The reality is that violence may very well happen despite extensive training, especially in life-threatening situations. While research often characterizes movements as primarily nonviolent or primarily violent, the reality is that "primarily nonviolent" campaigns frequently contain violence. About 85% of campaigns identified by the Nonviolent and Violent Campaigns and Outcomes (NAVCO) project include at least one riot.[22] Looking specifically at cases of democratic transitions, nearly 70% of transitions that were brought about by protest movements contained collective violence.[23] The NAVCO project itself shows that up to 75% of campaigns have a radical flank at some point.

What should be done when violence occurs? Understanding the reasons why violence happens is key. In particular, I advocate for paying close attention to the difference between spontaneous and planned collective violence. Starting with spontaneous collective violence, modern social science research is based on the premise that "rioting" emerges "as a rational way for the powerless and oppressed to fight back against disadvantage and discrimination."[24] Rather than actions of a senseless mob or criminal minds, violence by protestors is a symptom of deep and significant grievances. As King famously noted, "a riot is the language of the unheard."[25] Persistent injustice alone does not generate violence, and there are many studies dedicated to understanding what environmental conditions or immediate triggers translate grievances into collective violence. At times, specific events like murder or rape generate spontaneous violence. Repression is another of the most prevalent reasons why nonviolent discipline breaks down and collective violence emerges, as protesters often turn to violence when self-defense is required.[26]

In comparison to spontaneous collective action, radical flanks may be prompted by a sense that nonviolent methods are ineffective. If movements hit roadblocks and fail to make progress, violent alternatives become more attractive. Whether groups are faced with repressive policing, competition with other groups, or the activation of militant networks, violence may become a viable option in the face of lagging returns from nonviolent tactics.[27] The development of a radical flank, in other words, is linked to the activation of preexisting violent actors or the organizational reorientation of previously nonviolent groups; it happens over time, as groups interact with the state, competing organizations, and their own constituents. It is usually a strategic choice rather than a spontaneous eruption. Since radical flanks and planned collective action tend to be a choice, they do not fit within the context of maintaining nonviolent discipline. From the perspective of nonviolent action, the best approach is to avoid this purposeful adoption of violence when possible.

The relevant question is how leaders should respond to spontaneous violence if and when it occurs. When campaigns are unable to maintain non-violent discipline, a pragmatic approach takes advantage of this violence and the reasons *why* it emerges. Movement leaders can challenge political actors, pundits, and ordinary citizens to understand the grievances motivating individuals to act. In the spirit of "all attention is good attention," campaigns can use increased media coverage and discussion to their advantage by pointing to the reasons behind the demonstrators' collective violence wherever possible. Highlighting the use of police repression can encourage backlash. Stressing grievances can provide a platform to discuss the value of the campaign's goal. This approach, based in a desire to uphold moral actions and traditions, appreciates the sentiment behind violence and pays homage to the voices of the otherwise unheard.

The Black Lives Matters protests immediately following George Floyd's death serve as an example. As I saw in my own community and far beyond, the violence was described in decidedly negative terms, but also resonated with a public that is so often comfortable with, and in many ways complicit in, the pain suffered by so many for so long. It served as a call to take the accusations of the protesters seriously. At the height of the protest wave, a Pew Research poll found that nearly two-thirds of Americans supported the protests and said that long-standing issues related to the treatment of African Americans caused the protests.[28] In this case, the intensity of the situation may have translated into support for the movement and its goals.

Is Violence Justified?

Activists call for nonviolent discipline, but this requires protesters to risk repression while further restricting the campaign's tactical repertoire, potentially limiting their ability to achieve their overall goals. Given these costs, is violence justified? There are occasional benefits to violent tactics; however, it is hard to recommend the purposeful adoption of violence during nonviolent campaigns. Gains made by violent demonstrations during the U.S. civil rights movement were countered with long-term economic and political setbacks. Radical flanks may not have a strong negative impact on social movements, but they depress turnout and provide a justification for the use of harsh state repression. The conclusion from Haines, that positive radical flank effects are typically *unintended*, is particularly damning.[29] Nonviolent resistance is the stronger choice from both a moral and a strategic perspective for long-term success.

What about the short-term costs? I advocate for the continued use of training to lessen the impact of repression wherever possible, but I also recognize that training will go only so far in preventing injury and death. This is undoubtedly a risk undertaken by the protesters. The existing evidence suggests that movements are better able to succeed when nonviolent discipline can be maintained, making the overall benefits of maintaining nonviolent resistance a worthwhile endeavor. Even so, there is a good chance that violence will occur. When leaders are unable to maintain nonviolent discipline, a pragmatic approach takes advantage of the reasons *why* violence emerged.

There are many practical and moral consequences of violence, such that it should not be a purposeful strategy; however, if carefully done, violence can be bridled and used to bring attention, educate, and push the agenda forward. Violence is not the answer, but perhaps it can be part of the solution.

Notes

1. See, for example, Christian Davenport, "Multi-Dimensional Threat Perception and State Repression: An Inquiry into Why States Apply Negative Sanctions," *American Journal of Political Science* 39, no. 3 (1995), 683–713.
2. Martin Luther King, *Stride toward Freedom: The Montgomery Story* (Boston: Beacon Press, 1958), 85.
3. Erica Chenoweth and Maria Stephan, *Why Civil Resistance Works: The Strategic Logic of Nonviolent Conflict* (New York: Columbia University Press, 2011).
4. Jordi Muñoz and Eva Anduiza, "'If a Fight Starts, Watch the Crowd': The Effect of Violence on Popular Support for Social Movements," *Journal of Peace Research*, March 19, 2019, 1–14.
5. Chenoweth and Stephan, *Why Civil Resistance Works*, 119–120.
6. William J. Collins and Robert A. Margo, "The Economic Aftermath of the 1960s Riots in American Cities: Evidence from Property Values," *Journal of Economic History* 67, no. 4 (2007): 849–883.
7. Jamein P. Cunningham and Rob Gillezeau, "Don't Shoot! The Impact of Historical African American Protest on Police Killings of Civilians," *Journal of Quantitative Criminology* 37, no. 1 (2019): 1–34.

8. Jeremy D. Mayer, "Nixon Rides the Backlash to Victory: Racial Politics in the 1968 Presidential Campaign," *The Historian* 64, no. 2 (December 2001): 351–366.

9. "Dr. King Warns That Riots Might Bring Rightists' Rule," *New York Times*, February 18, 1968.

10. Steven I. Wilkinson, "Riots," *Annual Review of Political Science* 12, no. 1 (2009): 329–343.

11. Kadivar and Ketchley use the term "unarmed" collective violence, which is rather similar to "demonstrator" collective violence. There are many acts of collective violence that involve firearms, however, which generates questions about how to define "unarmed" actions. Instead, this term focuses on who organized the event. For more, see Mohammad Ali Kadivar and Neil Ketchley, "Sticks, Stones, and Molotov Cocktails: Unarmed Collective Violence and Democratization," *Socius* 4 (January 2018): 1–16.

12. The term "radical flank" is often used to describe a group that has a more radical goal or uses more radical tactics (not necessarily violence). Here, I am specifically referring to a group that uses violence and has the same overarching goals of the larger movement. Useful references for the concept of violence, riots, and radical flanks include Peter Turchin, "Dynamics of Political Instability in the United States, 1780–2010," *Journal of Peace Research* 49, no. 4 (July 2012): 577–591; Wilkinson, "Riots"; Erica Chenoweth and Kurt Schock, "Do Contemporaneous Armed Challenges Affect the Outcomes of Mass Nonviolent Campaigns?," *Mobilization: An International Quarterly* 20, no. 4 (December 2015): 427–451.

13. Kadivar and Ketchley, "Sticks, Stones, and Molotov Cocktails."

14. Sherry Hamby, Lawrence Grandpre, John Hope Bryant, and Noche Diaz, "When Is Rioting the Answer?," *Zócalo*, July 8, 2015, https://www.zocalopublicsquare.org/2015/07/08/when-is-rioting-the-answer/ideas/up-for-discussion/.

15. Chenoweth and Schock, "Do Contemporaneous Armed Challenges Affect the Outcomes of Mass Nonviolent Campaigns?"

16. Margherita Belgioioso, Stefano Costalli, and Kirstian Skrede Gleditsch, "Better the Devil You Know? How Fringe Terrorism Can Induce an Advantage for Moderate Nonviolent Campaigns," *Terrorism and Political Violence* 33, no. 3 (March 2019): 596–615.

17. Haines, Herbert H. *Black Radicals and the Civil Rights Mainstream, 1954–1970.* Knoxville, TN: University of Tennessee Press, 1995.

18. Haines, "Radical Flank Effects," 1, emphasis in original.

19. Kai Thaler, "Violence Is Sometimes the Answer," *Foreign Policy*, December 5, 2019, https://foreignpolicy.com/2019/12/05/hong-kong-protests-chile-bolivia-egypt-force-police-violence-is-sometimes-the-answer/.

20. Nadine Bloch, *Education and Training in Nonviolent Resistance* (Washington, D.C.: U.S. Institute of Peace, 2016).

21. Bloch, *Education and Training in Nonviolent Resistance*, 13.

22. Benjamin Case, "Riots as Civil Resistance: Rethinking the Dynamics of 'Nonviolent' Struggle," *Journal of Resistance Studies* 1, no. 4 (2018): 9–44.

23. Kadivar and Ketchley, "Sticks, Stones, and Molotov Cocktails."

24. Wilkinson, "Riots," 335.
25. Martin Luther King, "The Other America," speech, in The Radical King, ed. Cornell West, pp. 235–245, (United States: Beacon Press, 2015).
26. Jonathan Pinckney, *Making or Breaking Nonviolent Discipline in Civil Resistance Movements*, ICNC Monograph Series (Washington, D.C.: International Center on Nonviolent Conflict, 2016).
27. Donatella Della Porta, *Clandestine Political Violence* (Cambridge: Cambridge University Press, 2013).
28. Kim Parker, Juliana Menasce Horowitz, and Monica Anderson, "Majorities across Racial, Ethnic Groups Express Support for the Black Lives Matter Movement," Pew Research Center, Social & Demographic Trends Project, June 12, 2020, https://www.pewsocialtrends.org/2020/06/12/amid-protests-majorities-across-racial-and-ethnic-groups-express-support-for-the-black-lives-matter-movement/.
29. Haines, "Radical Flank Effects."

4

How Is It to Be Done?

Dilemmas of Prefigurative and Harm-Reduction Approaches to Social Movement Work

Ashley J. Bohrer

Over the course of a decade of activism, I have had the privilege of organizing with a variety of social movements for justice and liberation at the local, national, and international levels. I've worked with large NGOs with multimillion-dollar budgets and scrappy all-volunteer activist collectives without a dollar to their name. Across all of these experiences, I have participated in conversations and decisions about position and vision, but some of the most compelling dilemmas my comrades and I have faced have been less about *what* we believe and more about *how* we should bring about change. This isn't to say that we don't have debates to answer "What is to be done?," but the most generative, conflictual, and complicated decisions to navigate are often to answer "How is it to be done?" Engaging in the debates as an activist and reflecting on them as a scholar of social movements has revealed to me that one of the most important dilemmas that activists face is which framework to deploy in orienting our strategies for achieving lasting, impactful change.

While there are many dimensions to this strategic debate, in this chapter I would like to focus on one in particular: the divergence between prefigurative and harm-reduction approaches within social movements. By "prefigurative," I mean approaches to social justice work that strive to implement a future just society in the here and now; in its most stringent orientations, all decisions, processes, structures, and tactics activists pursue now must be consistent with the world they hope to bring about. By "harm-reduction approaches," I mean the orientation to political action that strives to reduce present suffering by whatever means available in the moment, even when those means might not reflect a perfect instantiation of political vision. The choice between these two approaches has been a constant in my experience

Ashley J. Bohrer, *How Is It to Be Done?* In: *Wicked Problems*. Edited by: Austin Choi-Fitzpatrick, Douglas Irvin-Erickson, and Ernesto Verdeja, Oxford University Press. © Oxford University Press 2022. DOI: 10.1093/oso/9780197632819.003.0005 of activist work across many different issues, organizations, and organizing structures.

To clarify the distinction between these positions, I find it helpful to think about them through two often counterposed maxims of organizing work that embody them. One comes from Audre Lorde: "The Master's Tools Will Never Dismantle the Master's House."[1] Evoking imagery of the plantation, Lorde powerfully stages the impossibility of relying on instruments of oppression in order to undo it. She calls attention to the tendency of movements, despite their best intentions, to replicate the very domination and oppression they seek to contest. In part, this unwitting replication is caused, Lorde tells us, by a reliance on the institutions, strategies, and discourses of power. But by mobilizing instruments of violence, Lorde says, we will never find our way out it. Thus, liberation politics must be prefigurative; that is, we need to act always in accordance with the highest version of our values or risk becoming the thing we hate.

Social movement writer and facilitator adrienne maree brown deepens this line of thinking into a full-scale model of social justice organizing. She argues that how activists do their work is just as important as their goals. In what she calls "emergent strategy,"[2] brown argues that the way we approach our work, even in small-scale collectives, will impact how we envision and implement larger-scale change. brown thus expands Lorde's insight about the dangers of replicating the expectations of power, extending the scope of the "master's tools" all the way down to individual interpersonal relations in activist spaces and the hundreds of micro decisions that activists and movements make. This perspective is rooted in the question: What really have we done in the end if we have reconfirmed, perhaps even strengthened, the illusion that the master's tools are all there is?

A competing maxim of social justice strategy comes from Malcolm X: "By Any Means Necessary."[3] On a certain reading of this perspective, all tactics, all institutions, all discourses are at least potentially available to social justice movements—even perhaps the master's tools. Here, the urgency of countering domination takes center stage; in situations of intense structural violence, expediency is a central value. Rather than a prefigurative approach to political work, this line of thinking foregrounds the necessity of using every tool available to us, because the dangers of not doing so are too great.

A modified version of this position asks whether oppression and domination have so saturated not only the field of power but even our own minds such that we may not have access to tools that are free from any relationship to the master. As decolonial theorist Serene Khader argues, resistance by definition occurs under nonideal conditions. For that reason, Khader asserts that in the contexts of feminist movements, for example, "practices that effectively increase gender justice in an unjust society may diverge from those that would exist in a society that was already just."[4] For Khader, evaluating strategies of resistance through their fidelity to prefigurative politics may skew our understanding about how resistance operates; she continues, "To claim that an exercise of power is resistant is to make a claim about its capacity to bring about diachronic change, not about whether we would expect to see it under conditions of gender justice."[5] Khader here highlights what in activist spaces is often called a "harm-reduction" philosophy: resistance strategies should be evaluated in terms of whether they overall reduce harm in a society, even if or when they may make use of problematic existing institutions, discourses, tactics, or tools. The Malcolm X–Khader approach warns that while some amount of introspection is necessary for social justice work, the Lorde-brown approach can sometimes direct all organizing energies inward, in a constant self-excoriation that can detract from mobilizing and organizing against the overarching structures of power. This perspective is rooted in the following questions: If we are the products of the master's tools, if we have been shaped and made by them, do we really have anything else? If we don't have anything else, can we not fight at all? How long do we have to work on ourselves before we can fight structures?

In what follows, I would like to raise some of the many considerations that activists consider in adopting one approach or the other. I counterpose these logics not in order to argue for one or the other but in order to illuminate what is different in them, what is both compelling and constraining. I seek neither synthesis between them nor to crown one over the other as the more ethical approach to doing social justice work. The consideration here is a rough sketch and is by no means exhaustive, but rather constitutes an entry point into the multifaceted dilemmas of each approach.

Prefigurative Politics

Perhaps the most compelling reason for adopting a prefigurative approach to politics is the strength of the political vision it offers. Rather than relying on the often pessimistic realism of "there is no alternative" philosophies, prefigurative politics can galvanize movements into seeing the radical possibilities of another, more just and liberated world. And rather than assume a kind of futuristic utopia, prefigurative politics suggests that, at least in some areas of our work, we can instantiate the world we want to see in the here and now. It offers a vision and a possibility for something other than piecemeal reforms or band-aids on the gaping wounds of contemporary injustice. Prefigurative politics is rooted in the possibility of radical action and therefore foregrounds the agency of both individuals and collective communities in bringing about lasting, deep, and substantive social change. The vision of this approach is deeply committed to the idea that we do not have to settle for the world as it is, that the fundamental rules of the games are open to direct challenge.

Prefigurative politics also open up an ethical position that refuses complicity. This refusal of complicity is not only a moral high ground but can have significant impacts on the choices of campaign targets, social movement rhetoric, and community building. It can thus open up the widest vision for social transformation and often refuses to leave behind some of the most marginalized groups in society. To be more concrete, many campaigns for social reform often focus on the most sympathetic members of marginalized groups in ways that can often reconfirm existing narratives of who is worthy. This debate came to a head in the Civil Rights Movement, for example. Months before Rosa Parks refused to give up her seat on a segregated bus, Claudette Colvin engaged in the same protest. However, because she was a fifteen-year-old single mother, civil rights organizations refused to make her the face of their antisegregation campaign. Even Parks once said of her case, "If the white press got ahold of that information, they would have [had] a field day. They'd call her a bad girl, and her case wouldn't have a chance."[6] In activist spaces, this kind of choice is often derided as a form of respectability politics, which plays into existing oppressive structures of power and unjust normative expectations that deem Colvin less worthy of liberation than Parks. A prefigurative approach would reject the splitting of oppressed populations into the "worthy" and "unworthy," both because all racialized people deserve liberation and because the marginalization of women of color's reproductive choices is a central part of structural white supremacy. The prefigurative position would say that one cannot dismantle the master's house (white supremacy) by utilizing the master's tools (white supremacist patriarchy). Strategically, the prefigurative approach proceeds with a simple, clear, and galvanizing set of political principles: all racial injustice must be uprooted, without exception.

This is one reason many activists believe that the prefigurative approach is in a stronger position to refuse the divide-and-conquer strategy that has been so central to contemporary forms of injustice. For this reason, it may be more capable of instantiating intersectional critiques that show the full extent of existing injustices. In the Colvin example, the prefigurative approach can also show how the gender politics of reproductive control is a central tool of white supremacy; in so doing, it can ground a more comprehensive analysis of what is wrong in the world and what would need to be different in an emancipated society.

This version of politics also strengthens the connection between abstract systems of power and smaller-scale actions by groups and individuals. By radicalizing a version of the oft-deployed slogan "The personal is political," it can open up avenues for individual and collective transformation in line with visions of emancipation. Recognizing the ways that individual actions can either reaffirm or refuse larger systems of power, it can also lend a concreteness to analyses of injustice that are demobilizing barriers for social justice action. If systems of power are so vast and so well-entrenched, then uprooting them can seem like an impossible and never-ending task. But when the prefigurative approach can compellingly link these abstract structures to concrete practices, social justice work can feel more tangible.

However, like any approach, prefigurative politics can also harbor certain difficulties for social justice work. The strength and comprehensiveness of vision, while it can sometimes make politics feel concretely interpersonal, can also make "winning" feel impossible. In meditating so unflinchingly on the multiple levels of injustice that one must simultaneously fight—in oneself, in one's community, and in the world—this approach can sometimes make activists feel as if any gains are so minimal as to be completely unimportant. It can also be just plain exhausting to critique everyone all the time. The comfortable certainty of standing on the right side of history often becomes destabilized in prefigurative organizations, as we come into constant consciousness of the ways that we may be constantly reproducing the very harms we intend to fight.

Prefigurative politics can also at times make forming cross-organization alliances more difficult. In some versions of this approach, any organization that has failed to deeply interrogate each of its own practices, stances, and discourses may be dismissed as potential movement partners. It can thus be difficult for prefigurative organizations to galvanize newly radicalized people, who may not see every layered dimension of injustice, in themselves or in the world. But this can also lead to a politics of purity that prevents us from forming alliances across differences of social position, proximity to power, or ultimate political vision, leading to intractable left sectarianism. This is not to say that small groups cannot be powerful or impactful, but they certainly face deep impediments to developing *mass* movements and *mass* struggle. It can therefore be a more comfortable approach for seasoned, longtime activists rather than new activists, which may negatively impact the ability to build truly mass movements and engage newer or more moderate participants. Additionally, some groups that are deeply rooted in prefigurative politics may also be less receptive to working in coalition with those whose visions or tactics differ from their own (in order to prevent complicity), and for that reason may tend to constitute more ideologically homogeneous groups. In itself, this is not necessarily a problem, as many small groups with shared vision can make a lasting impact on the world; it may, however, prevent some of the fruitful cross-pollination of ideas, tactics, and approaches that larger, more ideologically heterogeneous groups benefit from in the context of mass movement spaces.

In some iterations of prefigurative politics, the necessity of constantly interrogating the *how* of politics can lead groups to turn excessively inward, focusing on the important projects of unlearning and reshaping internal organizing dynamics and rooting out any vestiges of oppressive practices within the ranks. This is of course a worthy and necessary dimension of political activism, but it can also reorient a group's focus away from changing external conditions into a hyperfocus on the activists themselves or to internal dynamics that may leave less time for confronting external deployments of power.

The prefigurative approach to politics thus contains both opportunities and challenges for social movements on multiple levels. Overall, the deep commitment to instantiating a new, more just vision of the world remains one of the most powerful assets of prefigurationist approaches to social justice, even if and when implementing that vision may be rife with complications and dilemmas of its own.

Harm-Reduction Politics

"Harm reduction" as a term came out of radical approaches in public health communities, specifically around issues like substance use, sex work, and other stigmatized practices. Over the years, however, the language of harm reduction has become more common in activist spaces far beyond this initial context, looking at everything from passing political reforms to voting as a method for radical activists to engage with powerful institutions. From a harm-reduction perspective, working with or inside oppressive institutions of power is a negotiated process that centers how and whether certain tactics can reduce the pain, suffering, and violence that oppressed communities experience in the immediate term.

Harm reduction's greatest strength may be its actionable immediacy. Because the paramount value of harm reduction is present effects rather than long-term vision, its ultimate fidelity is to the concrete circumstances in real people's lives. This means that the calculus deployed by harm reductionists is less "Is this course of action good or bad?" in the abstract and more "Does this course of action lead to better or worse effects than the status quo?" In grounding action in an evaluation of "better or worse," it opens the pathway for more limited reform strategies that can have substantive effects in the short term.

Harm-reduction approaches can thus use a wide range of resources for engaging in social justice work, even when those resources may have complicated and unsavory histories that activists may ultimately condemn. One of the ways that harm reduction has formed a galvanizing flashpoint in recent years among radical activists is on the question of voting. In many activist communities that view the government as an essentially oligarchical, settler colonial, patriarchal, white supremacist institution, voting can be seen as complicity with that system, strengthening the government's narrative of self-justification. Harm reduction has taken another approach: while recognizing that the U.S. government is indeed the source of a significant amount of oppressive and exploitative harm, voting for particular policies or candidates can make a difference in the lived experiences of those who are harmed by the system. It may make a significant difference in the lives of many racialized communities, for example, if drug possession is decriminalized. A harm-reduction approach would argue in this case for engaging an oppressive institution in order to reduce the amount of harm that institution will likely perpetuate.

This selective engagement with institutions of power can also lend an extraordinary concreteness to social justice demands; rather than aiming, in the short term, for the abolition of entire, complicated systems, activists can mobilize around particular policies with concrete evidence for how real lives would be affected by this proposed change. For this reason, movement gains may be easier to argue for and thus more accessible to groups of people outside of the radical left. In order to be effective, harm-reduction approaches can thus often sidestep larger and more abstract conversations about systems of domination in order to deliver concrete wins.

In focusing on smaller and more specific aims, harm-reduction approaches can also hold more heterogeneity of vision within movement spaces, and are therefore well-positioned to be effective at building broad-based mass coalitions for targeted reform. Because they can often more easily reach nonactivists, harm-reduction philosophies can also be important bridges into more radical social movement spaces.

But the philosophy of harm reduction can also present deep and lasting challenges. The harm-reduction calculus implicitly suggests that the goal is reducing—not eliminating—harm. It can also be difficult to determine *whose* harm will be prioritized, especially in situations where group interests may be opposed to one another.

Harm-reduction approaches can also be much more easily co-opted by powerful institutions for their own ends. In working with and alongside institutions of power, activists may face immense pressure to dampen their rhetoric or to accept compromises that reproduce harm in other ways. As many critics of police brutality have shown, some of the most vaunted harm-reduction approaches, like bias training and body cams, have actually not led to any significant reduction in police-perpetrated violence or extrajudicial murder, while at the same time police departments are able to demonstrate their implementation of these reforms in order to strengthen their legitimacy in the eyes of the general public and lawmakers. Under such conditions, the ultimate goal of harm reduction—in this case, the reduction of police violence—can actually lead to a further insulation from critique that allows the police to continue deadly practices. Especially when reforms become institutionalized by the state or other unwieldy institutions that lack broad-based accountability, policies that were initially intended to reduce harm are taken out of activists' hands and placed where the definition of harm and its modes of assessment may be implemented in ways that significantly diverge from their original activist visions.

Overall, the harm-reduction approach to social movement work is most compelling in its commitment to being rooted in the real, lived experiences of violence under the contemporary organization of society and the deep and abiding commitment to intervening in that harm in the immediate moment.

The dilemmas that arise out of this approach should always be taken in light of this commitment.

The Dilemmas and Promise of a Both-And Approach

Of course, actually existing individuals and organizations rarely fall neatly into one perspective or the other, and in any case, each of these approaches contains elements of the other. Many elements of harm reduction are also prefigurative: harm-reduction approaches often foreground agency, empathy, care for others, empowerment, bodily autonomy, and other values that certainly are not structuring conditions of our world. At its base prefigurative politics is rooted in the desire and the imperative to prevent the reproduction of harm, that is, to reduce harm to zero. Many activists and organizations try to model a kind of blended approach; as legal scholar Mari Matsuda put it in "When the First Quail Calls":

> On the one hand, they respond as legal realists, aware of the historical abuse of law to sustain existing conditions of domination. Unlike the postmodern critics of the left, however, outsiders, including feminists and people of color, have embraced legalism as a tool of necessity, making legal consciousness their own in order to attack injustice. Thus, to the feminist lawyer faced with pregnant teenagers seeking abortions, it would be absurd to reject the use of an elitist legal system, or the use of the concept of rights, when such use is necessary to meet the immediate needs of her client. There are times to stand outside the courtroom door and say "this procedure is a farce, the legal system is corrupt, justice will never prevail in this land as long as privilege rules in the courtroom." There are times to stand inside the courtroom and say "this is a nation of laws, laws recognizing fundamental values of rights, equality and personhood." Sometimes, as Angela Davis did, there is a need to make both speeches in one day.[7]

Matsuda highlights the ways that some activists and movements have pursued a "both-and" strategy rather than exclusively functioning within one domain or the other. The dilemma, however, for activists interested in the both-and approach is in determining not only the overall balance between them but also when one might be more appropriate than the other.

All that is to say that the attempt to find a "middle ground" between these approaches may also be rife with its own dilemmas. Indeed, I have often found that when activist communities attempt to pursue prefigurative and harm-reductionist approaches simultaneously, they do not simply harness the strengths of each; they often also take up the weaknesses of both approaches simultaneously, with the additional dilemmas of selecting how and when and where to choose one over the other in any particular instance.

Movements are often strongest when they contain organizations that adhere to both strategies. The cross-approach critique that can often arise in vibrant and heterogeneous movements is itself a deeply important element of activist work; it is by reminding each other of the dangers in our chosen approaches that we often see our own work anew and can engage in necessary projects of reevaluation and reorientation. Fredric Jameson once wrote—in a *completely* different context—"The two codes must criticize each other, must systematically be translated back and forth into one another in a ceaseless alternation, which foregrounds what each code cannot fully say as much as what it can."[8] While individual activists and organizations must make specific choices between these approaches to how they accomplish their goals, Jameson's insight foregrounds how being in community with those who make different, even counterposed strategic decisions can itself be an important service to the vitality of social movements.

The reality of engaging with radical social justice work is that, despite the obviousness of many movement slogans, the embodied practices of engagement often present deep and lasting dilemmas, which are neither easily avoided nor resolved. In many cases, they may seem intractable. My aim in raising these concerns is not to condemn movements that make use of one approach or the other or their complicated synthesis, but rather is based in the conviction that understanding the potential dangers of various movement strategies is crucial to being able to evaluate them and to really *choose* them. Understanding the pitfalls of our chosen orientations allows us to be vigilant about them. Liberation movements can always benefit from clearer analyses of the limitations of their chosen orientations; being self-critical of our tactical choices is itself an ethical orientation to liberatory praxis. It remains vital to recognize and reflect on intramovement debates not only about *what* is to be done but, crucially, *how* it should be done.

Notes

1. Audre Lorde, *Sister/Outsider: Essays and Speeches* (Freedom, CA: Crossing Press, 1984).
2. adrienne maree brown, *Emergent Strategy: Shaping Change, Changing Worlds* (Chico, CA: AK Press, 2017).
3. Malcolm X, *By Any Means Necessary: Malcolm X's Speeches and Writings* (New York: Pathfinder Press, 1992).
4. Serene Khader, *Decolonizing Universalism: A Transnational Feminist Ethic* (New York: Oxford University Press, 2019), 130.
5. Khader, *Decolonizing Universalism*, 133.
6. Guardian Staff, "Civil Rights Heroine Claudette Colvin," *The Guardian*, December 15, 2000, https://www.theguardian.com/theguardian/2000/dec/16/weekend7.weekend12.
7. Matsuda, Mari. "When the First Quail Calls: Multiple Consciousness as Jurisprudential Method." Women's Rights Law Reporter *11*, nos. 1 (1989): 7–10.
8. Fredric Jameson, *Valences of the Dialectic* (London: Verso, 2009), 47.

SECTION II
LEADERSHIP AND ORGANIZATIONS

The previous section discussed the utility and limitations of strategic violent and nonviolent action, as well as dilemmas over whether one should work within existing systems of domination to effect change, or whether it is better to reject those systems in favor of a radically principled approach that may nevertheless lose some practical efficacy. In this section we move to a different set of challenges, namely the dilemmas that arise around leadership and organizational behavior. Who is entitled to speak for others, and on what grounds? What are the strengths and trade-offs in working with allies, and how should leaders handle valuable external support that may nevertheless threaten to redefine an organization's agenda and priorities or even undermine local peacebuilding autonomy and capacity?

Leaders and organizations often find themselves having to make difficult compromises in order to bring about practical change, but over time this can risk co-optation, tokenization, or rationalizations that shield them from the difficult task of reflecting on the uncomfortable consequences of their actions. Here, contributors examine these issues around leadership, alliance building, and ethical reflection through a series of provocative chapters covering cases around the world. Some of the pieces are highly personal and situated in concrete struggles, revealing the profound challenges that peacebuilders confront, while other chapters take a broader analytical perspective to think more systematically about these questions. All, however, seek to sharpen the dilemmas and choices at stake and decide how to navigate them while keeping faithful to one's basic ethical principles.

5

The Paradox of Survivor Leadership

Minh Dang

Four years into my advocacy as a survivor of human trafficking, I was invited to speak at a high-profile event in Washington, D.C. Not wanting to go alone, I invited a friend and fellow survivor to join me. I knew that she was interested in becoming a survivor leader and that this event would give her some insight into the world of survivor advocacy. The event went well, and we returned to California, where we lived. Shortly after the trip, my friend asked if we could meet to debrief the event. I happily agreed and met her during her lunch break at a café in downtown Oakland. She was already settled at a table, and I made my way through the maze of mismatched chairs. After some initial conversation and checking in, she said, "I'm wondering if I can give you some feedback."

"Yea, sure," I replied.

"When we were in D.C., you seemed to treat me and the other survivor who was there like you were better than us. You treated us like you are the bottom b****,[1] and I'm wondering if you are just in this for the fame and money."

About eight years into my advocacy, a journalist reached out with the hope of interviewing me about "my story." I sent him my standard reply, which emphasized that I'm happy to talk about my work to promote survivor leadership but that I will not share the details of being exploited. In his reply the journalist highlighted that his editor and audience find a "personal story" easier to connect to. I sent him a tip sheet that I wrote for journalists about how to interview survivors of human trafficking, and an explicit list of questions that I would not answer. If he agreed to avoid those questions, I would be happy to speak with him. I also mentioned that my story in advocacy *is* personal, but it won't reveal the gory details of my trauma.

The journalist agreed, we set up a time to speak, and the interview was conducted over the phone. A few questions into the interview he said,

Minh Dang, *The Paradox of Survivor Leadership* In: *Wicked Problems*. Edited by: Austin Choi-Fitzpatrick, Douglas Irvin-Erickson, and Ernesto Verdeja, Oxford University Press. © Oxford University Press 2022. DOI: 10.1093/oso/9780197632819.003.0006

"Elsewhere, you speak about the fact that you were sexually abused regularly. Can you tell me how many times a day that occurred?"

The Paradox

As I hope you can imagine without my saying it, I was furious after each of these interactions. The journalist received some sharp words from me and I ended the interview. I emailed the journalist and his editor that I was withdrawing my consent to have my interview published. Unfortunately, I was not able to rebuke my friend's comments as quickly, and I entertained them internally for months. Many years have passed, and we are no longer friends. And, for the record, while my travel costs were paid, I never received nor requested a speaking fee for that engagement.

Experiences like these are sadly too familiar to me and to a handful of survivor leaders who have confided in me. The experiences outside of the survivor community—with journalists, for example—are painful but a bit easier to handle. Experiences within the survivor community—with friends, for example—can cut deep wounds. Both experiences can drive one away from survivor leadership and deter new survivor voices from entering the field. However, without survivor voices, many survivors suffer at the hands of well-intentioned, but ineffective, programs and policies. Staying silent means sitting by and watching professionals, who are distanced from the lived reality of human trafficking, stumble and falter their way through prevention and intervention efforts that could easily benefit from lived experience expertise. As survivor leaders, we feel a double bind: How can we sit by and not lend our voices, when we believe our insights can benefit other survivors from suffering as much as we have? And yet, is speaking up worth the cost to our own lives and the extra burden that we will carry? Here we find a central dilemma: Is securing a leadership position a crucial step in gaining power for the movement, or is it a critical blow to one's ability to represent one's community?

This chapter addresses the wicked problem of ethical representation in social movements. The double bind that I present is paradigmatic in social movements that seek to incorporate the voices of people most affected by an injustice. My insights into this problem come from my work as a community organizer in antitrafficking since 2003. My advocacy as a survivor of human trafficking has not focused solely on my own narrative. This was the case in my early years in the field, but I witnessed minimal systemic change occurring as the result of my peers and I sharing our trauma narratives. Collectively, we began to make policy recommendations, consulted with nongovernmental organizations (NGOs) to improve service provision, pursued higher degrees, participated in research, and launched our own organizations.

One of these organizations is the one I currently lead, Survivor Alliance, an NGO that unites and empowers survivors of slavery and human trafficking around the world. As of 2021, we are a network of over 350 people from twenty-four countries who self-identify as survivors of slavery or human trafficking, based on UN definitions. To be clear, survivors are people who have lived experience of human trafficking or slavery but who are no longer being exploited. Our mission is to unite and empower survivors in order to buoy ourselves and build capacity to become leaders in the antitrafficking movement. We are one of the few antitrafficking organizations that are survivor-led and focused on leadership development and capacity building rather than direct service provision. Survivor Alliance facilitates opportunities for survivors to engage publicly or anonymously in antitrafficking work via consulting, research, focus groups, peer support groups, and advocacy.

My work at Survivor Alliance has highlighted the ongoing costs to survivors who want to share the insights gained from lived experience. Tokenization and dehumanization are among these costs and resemble similar costs that other marginalized communities have faced when seeking justice. Whether social movements aim to elevate undocumented migrants, sexual assault survivors, residents of an impoverished neighborhood, or people with mental health disabilities, they are keenly aware that the voices of the oppressed are required for the movement to succeed. And yet the additional costs borne by the people who lend their voices and share the expertise gained from their experience of injustice are either ignored or accepted as necessary and unavoidable. We may be tempted to ask: *How do we incorporate the voices of people with lived experience, while preventing or at least mitigating additional injustices and suffering?* While this is a relevant and useful question, the questions that will result in more systematic change are these: *How do social movements develop, and how can they be conceived by and for the people who have suffered most? If movements have already begun mobilizing, how do we uproot the foundation of exclusion from the core infrastructure of movement leadership?*

Nothing about Us, without Us

Gifted to the world by the Disability Rights Movement, the phrase "Nothing about us, without us" succinctly summarizes the need for voices of those impacted by oppression. Disabled activists were calling for inclusion in the policies and practices that would apply to them. A simple concept has remained a radical stance, and social movements still struggle to heed these words. These words highlight the double bind that survivors of human trafficking face. The conditions that lead a community of people to call for representation inherently mean that exclusion is already endemic. Request for inclusion means a request for representation in a system that is already designed for exclusion. James I. Charlton in his seminal work, *Nothing about Us without Us*, speaks eloquently about this paradox:

> A fundamental paradox confronting the disability rights movement is that the progress of people with disabilities is contingent on significant economic development (the accumulation and expansion of capital) and, correspondingly, the emergence of (more) modern ideas about disability (the influence of capital) and, at the same time, the development of a movement that insists on social justice and equality (the restriction of capital) and an epistemological break with the dominant ideology (the rejection of capital).[2]

The economic and sociopolitical conditions that construct oppression are the same conditions within which people who are oppressed must seek to gain credibility, all while desiring to dismantle those very conditions. The tension between prefigurative and harm-reduction approaches to social justice, detailed by Ashley Bohrer's chapter in this volume, is on display in this situation. For survivors of human trafficking, the conditions that created their exploitation are the same conditions that create their exclusion from the antitrafficking movement. Specifically, in order for a survivor of human trafficking to participate in the movement as it stands, they must be able to earn an income from their antitrafficking work or finance their own time, travel, and expenses. Yet the fact of being a survivor of human trafficking means that the person experienced economic exploitation and psychological trauma that can impact daily functioning and the capacity to participate competitively in the capitalist market. Survivors of human trafficking must seek change within current institutions in order to have enough financial capital to even consider creating realities outside of the current exploitative situation.

Dehumanization and tokenization are produced when inclusion is prioritized in a flawed system. Dehumanization occurs when dominant groups treat people with lived experiences as essentialized identities. This happens when the oppressed are viewed only through the lens of their oppression. People who start out being known as a survivor of human trafficking, a mental health service user, or a disabled person often have a difficult time shaking that label despite an accumulation of work and professional experiences. Survivors are then faced with another bind: to hide their experiences of oppression or to reveal them at the risk of being known for nothing else.

Dehumanization also occurs when some or all lived experience experts are put on pedestals; then a small group of unique and charismatic leaders are sought out. Those who fit the stereotypical victim narrative, who are educated and articulate, and who have the privilege of being safe enough to be public about their stories of human trafficking are the leaders who grab the attention of funders, news media, NGOs, and the general public. These individuals are held up as gold-standard survivors, and others are compared to them. Survivors who don't meet this gold standard are often dubbed "too emotional," "unprofessional," or "not captivating." This communicates that there is a particular presentation that survivors must fit into before being valued and listened to.

Pedestalization is interlinked with tokenization. Inclusion of one "special" representative becomes a marker for dominant groups to indicate that they have done their inclusion work. Asked to speak for all, much like ethnic minorities are asked to speak for their entire group, individuals are treated as envoys for a diverse and complex population. Additionally, single narratives become co-opted, transformed for the use of social movements to seek wider public approval and to raise funds for organizations. Often, the result is the socially constructed stereotype of the "perfect victim," someone with whom the dominant group can easily empathize. Bohrer highlights this challenge as well; a harm-reduction approach intensifies the distinction between people who are "worthy" and "unworthy" of humane treatment. People with lived experiences that differ from this single narrative must expend additional energy to dispel this stereotypical image, often requiring them to share difficult personal narratives to demonstrate the falsehood of the stereotype.

In seeking inclusion from a predetermined, oppressive system, people with lived experience face additional consequences from the process of seeking inclusion. Addressing these consequences, or costs, as I refer to them, is an unfortunate burden that people with lived experience carry when they decide to share their experiences of oppression. The vignettes shared at the beginning of this chapter are only a window into these costs. People with lived experiences who bear these costs face emotional exhaustion and potential burnout. Since social movements need the lived experiences of people who have suffered oppression, how do movements sustain ethical representation?

Discussion

A key social movement debate related to this question is whether the best approach is to seek incremental change or revolution of the system writ large.[3] My experience suggests that both strategies are important from the perspective of well-being for people who are oppressed. Incremental change is required to improve immediate and short-term living conditions, but revolution of the system is required for long-term, intergenerational change. COVID-19 provides an example of this. Incremental change efforts over time created better conditions for some survivors of human trafficking to improve their life circumstances. These changes often occurred with the support of nonsurvivor allies who provide grant funding, run direct service programs, author legislation, implement policies, and/or conduct research. Working with allies is helpful for pushing any existing political organization toward greater alignment and sensitivity to the lived experiences of those they are meant to serve. During the global pandemic, many incremental changes for survivors were quickly reversed and put them at high risk of re-exploitation or other dire circumstances. Existing working relationships with allies enabled immediate survivor advocacy and rapid implementation of stopgap measures to reduce the negative impact of the coronavirus on survivors. Survivors who were already positioned to work with allies had greater access to information, capital, and leverage with decision-makers.

Incremental change is also important for political education and capacity building within communities that are oppressed. A core element of grassroots organizing is to fight for changes that are concrete and that materially improve people's living conditions. As Charlton emphasized, people who are oppressed typically need "significant economic development" to have enough capital to survive and to organize their communities. Campaigns for a livable wage, for enforcing equal-pay laws, and even for equal representation within social movements can achieve real-life gains for survivors. Participating in these campaigns, in the education and skill building required to enact them, enables people to recognize their own power, build community with others, and have a purpose—all of which are documented elements of psychological well-being.[4] When people who are oppressed fight for changes in their lives, it requires a shift in self-concept and a shift toward what C. Wright Mills coined "the sociological imagination," an understanding that personal problems are connected to social issues.[5] At Survivor Alliance, I regularly witness survivors of human trafficking who develop a sense of empowerment by advocating for their own rights and who begin to shed some of the self-blame that is a pervasive consequence of interpersonal violence. Additionally, fighting for incremental change allows survivors to learn about the sociopolitical systems that influence their lives and how to influence the levers of change in the dominant social structure.

Simultaneous with incremental change, we must begin to ask: *How do social movements start from and remain in the control of the people with lived experience of that injustice?* At Survivor Alliance, we believe that we need a critical mass of survivor leaders who can sustain a self-defined movement built on our own principles and with our own agenda—the one we have always been fighting for—a social and political system constructed in such a way that *every* human being is treated with dignity and no one can exploit another. The work of Survivor Alliance is rooted heavily in the traditions of liberation theology, made prominent by Paolo Freire's *Pedagogy of the Oppressed.*[6] Less well known is liberation social psychology (LSP), developed by Latin American scholar Ignacio Martín-Baró. LSP was born out of a critique of European and American psychology traditions, which focused solely on healing the individual. LSP approaches individual healing as intricately connected to the pursuit of collective social justice.[7] The core idea is that we cannot help individuals, ourselves included, to heal and then require us/them to continue existing in the same sociopolitical system that created the injury in the first place.

People with lived experience of any oppression need to have their own spaces to interact in alternative modalities from the oppressive dominant norm. In this space, like the consciousness-raising groups of the women's liberation movement,[8] people can heal from the wounds of oppression and organize for collective political action. It is also in this space that survivors will determine their own community gatekeeping methods and structures. Rather than a dominant group deciding whose voice is relevant or valid, the community itself will determine its spokespeople and hold them to account. A community of people who operate outside the dominant norm (as much as possible) can also develop community capacity to build and maintain solutions for themselves, in addition to increasing the availability of lived experience experts.

In other words, survivor capacity building enables survivors to focus on both incremental change and systems change. Internal capacity building increases the supply of lived experience experts who are available to *and interested in* consulting with the mainstream social movement. From a foundation of community life and community strength, those who have the emotional bandwidth, economic security, necessary skills, and community support can engage with the dominant system to seek incremental change. In this way, mainstream movements have fewer opportunities to tokenize survivors and cannot hide behind a thinly veiled victim-blaming stance of "There is no one with lived experience willing to talk to us." Simultaneously, survivors can trial community-based interventions and hold them up in comparison to or, at a minimum, as an alternative to mainstream interventions.

Conclusion

Lived experience expertise is essential to every social movement. By offering a specific standpoint epistemology, a way of knowing that is rooted in a social position and identity, people with lived experience of a social injustice provide the necessary insight and urgency needed for social change.[9] Even the most astute and empathetic allies will lack the unique perspective that comes from lived experience of human trafficking. As a cisgender woman, I will never know what it is like to be a transgender woman. My knowledge extends as far as my ability to empathize and educate myself about the experiences of transwomen. Although my voice may be important to support social justice for people who identify as transgender, it is still a *different* voice, a voice that comes from someone who has not experienced the nuances of living as someone whose gender is marginalized. The voices of many transgender people are needed to understand the diversity of experiences. Similarly, the voices of many survivors of human trafficking are needed for the antitrafficking field to understand the diversity of survivors' experiences. The variation enables novel contributions to programs and policies. Without these unique standpoints, the antitrafficking field squanders the opportunity to access innovative, context-specific, culturally informed knowledge from survivors.

In my experience, bringing survivor voices to antitrafficking efforts that are led and curated by the dominant social movement can have the effect of pushing the movement toward more survivor-centered priorities. Engaging with nonsurvivor allies can lead to concrete improvements in survivors' lives. Sharing our voices can also help us make meaning of our own experiences and empathize with others.[10] Many survivors of human trafficking want to have our voices become part of the larger movement and therefore continue to engage with it. We want people to know what exploitation and life *after exploitation* is really like, beyond the Hollywood version of unique cases and sensationalized media clips. Untold and unaddressed stories of trauma within a person's head and heart can even lead to long-term health consequences.[11] Additionally, many survivors want to share our lived experiences if it might stem the tide of violence and prevent others from experiencing the trauma that we have.

On the other hand, survivor voices cannot withstand ongoing tokenization in the mainstream movement. Survivor involvement cannot be carried on the shoulders of an elite few, as those individuals will burn out. We must prevent or at least mitigate the costs of systemic change that inevitably fall to movement incumbents. Survivors must continue to care for and lift up one another, showing up for our peers who seek incremental change in the mainstream movement and provide them a holding space to which they can return. We must build community strength and wealth to enable independent interventions that do not rely on mainstream buy-in. By doing so, we can defend against co-optation and avoid becoming enamored with wealth or status only to sacrifice our greater mission.

Ethical representation of survivors of human trafficking, and arguably survivors of violence and trauma, needs a systemic shift. This is an agenda that welcomes allies. But in order to get there, our allies must proactively share power, deconstruct and withdraw their investment in the status quo, and engage in the creative process of community building. This is not an easy task for allies, a challenge Daniel Myers explores in more depth in the next chapter. Allies must also be willing to shield people with lived experience from constantly justifying their worth and elevate people with lived experience at all opportunities. We invite you to ask and act strategic, systems-shifting questions. Instead of "What does ideal representation look like?," let's ask "What does a movement led by survivors look like?" No longer is it "nothing about us without us." It is time for social movements to be *by us and for us.*

Notes

1. A role in commercial sexual exploitation that is associated with a woman who is forced to be the most loyal to a perpetrator, engage as if she is actively in charge of the business, and facilitate transactions that exploit other victims.
2. James I. Charlton, *Nothing about Us without Us: Disability Oppression and Empowerment* (Berkeley: University of California Press, 1998), 154.
3. Anton Törnberg, "Combining Transition Studies and Social Movement Theory: Towards a New Research Agenda," *Theory and Society* 47, no. 3 (June 2018): 381–408, https://doi.org/10.1007/s11186-018-9318-6.
4. T. Coggins, "Translating Wellbeing Policy, Theory and Evidence into Practice," in *Wellbeing, Recovery and Mental Health*, ed. Mike Slade, Lindsey Oades, and Aaron Jarden (Cambridge: Cambridge University Press, 2017), 231–244.
5. C. Wright Mills, *The Sociological Imagination* (1959; New York: Oxford University Press, 2000).
6. Paulo Freire, *Pedagogy of the Oppressed* (New York: Herder and Herder, 1970).
7. Mark Burton and Carolyn Kagan, "Liberation Social Psychology: Learning from Latin America," *Journal of Community & Applied Social Psychology* 15, no. 1 (2004): 63–78.
8. Naomi B. Rosenthal, "Consciousness Raising: From Revolution to Re-Evaluation," *Psychology Women Quarterly* 8, no. 4 (June 1984): 309–326.
9. Sandra Harding, "Rethinking Standpoint Epistemology: What Is 'Strong Objectivity'?," *Centennial Review* 36, no. 3 (Fall 1992): 437–470.
10. Michael White and David Epston, *Narrative Means to Therapeutic Ends* (New York: Norton, 1990).
11. Robert F. Anda, Vincent J. Felitti, J. Douglas Bremner, John D. Walker, Charles Whitfield, Bruce D. Perry, Shanta R. Dube, and Wayne H. Giles, "The Enduring Effects of Abuse and Related Adverse Experiences in Childhood: A Convergence of Evidence from Neurobiology and Epidemiology," *European Archives of Psychiatry Clinical Neuroscience* 256, no. 3 (2006): 174–186, https://doi.org/10.1007/s00406-005-0624-4.

6

Allies Out Front

Dilemmas of Leadership

Daniel J. Myers

In the early part of 1987, while I was a student at Ohio State University and a resident advisor in one of the dorms, I took my first unsteady, even clumsy steps into being an activist for LGBTQ rights. Those beginnings were nothing dramatic—I wore some T-shirts, attached a pink triangle button to my fanny pack (it was the '80s after all!), and got into conversations with other students and the men on my floor about gay rights and tolerance.

In the next couple of years, as I progressed through graduate school, my commitment grew. By the time I was a full-time student affairs staff member in 1990, I had adopted what we then called "GLB" issues as a primary social justice cause. That growing dedication from my straight self was brought into full flower through my deep friendship with two other staff members working in student affairs, both gay men. Those relationships remain the most important formative influences on my activism and my thinking about being an ally—and being an ally in leadership.

"Ally in leadership." That's a loaded phrase and a lot to unpack—way more than I imagined when I started on my activist journey. Some activists welcome allies into leadership positions in social movement organizations and even laud those who land there, but others would say that allies just don't belong in leadership roles. That contradiction is powerful for a number of reasons, and it can make ally leadership challenging and uncomfortable. But it is also an incredibly important tension, and if you don't feel it as an ally leader you probably aren't paying enough attention—and maybe you shouldn't be in the role you're in.

Daniel J. Myers, *Allies Out Front* In: *Wicked Problems*. Edited by: Austin Choi-Fitzpatrick, Douglas Irvin-Erickson, and Ernesto Verdeja, Oxford University Press. © Oxford University Press 2022.
DOI: 10.1093/oso/9780197632819.003.0007

The "Minority Issues Committee"

Before I go further, let me introduce you to the other characters in this drama, my two friends, Robert and Jeff. Robert was, at least by Columbus, Ohio, standards of the time, a radical. He helped found the local Queer Nation chapter and had been active in ACT/UP (the incubator for Queer Nation). Robert advocated (as did Queer Nation) for selective outing and thought just about everyone should out themselves. Robert was about as far from closeted as you could imagine, and his introduction of himself often bordered on "Hi, I'm Robert. I'm gay, and you can either love me for it or get over yourself!" Most often, people loved him, but if not, he wasn't taking prisoners. Jeff had a lighter touch. I think it would be fair to say he was selectively closeted—he was open with some of his colleagues, and many of us knew his partner. He was deeply enmeshed in the gay community in Columbus but was critical of many of his friends who were not politically active.

Despite the political alignment between Robert and Jeff, and the fact that they were two of my very best friends, they were not close and did not see eye to eye. Indeed, while they both coached me in my growing awareness and activism, I often felt like a middle child negotiating between two estranged parents. Robert was too radical for Jeff, and the Queer Nation attitude toward outing vexed Jeff. Jeff was too quiet for Robert, who felt Jeff did not do enough to protect other gay and lesbian staff members or push the agenda forward at Ohio State. Both, however, appreciated my commitment to LGBTQ issues and the public stance I took with other staff, students, and the administration. But they did not equally appreciate, as it turned out, my role in leadership activities.

Around that time, Ohio State had put together a group called the Minority Issues Committee. The Minority Issues Committee was an umbrella made up of the nine or ten chairpersons of a series of other "issues" committees, such as the Hispanic Issues Committee, the Women's Issues Committee, the Disability Issues Committee, and so forth. One year, Jeff was asked to be the chair of the GLB Issues Committee. He didn't want to take on the whole thing alone, so he asked me to co-chair with him. In addition to bringing more person power to the operation, he thought that an open, vocal ally would help us to recruit and activate more allies. Furthermore, Jeff, while a gregarious and loquacious presence, was not the most organized person, and we both knew that I would help produce structure and traction for the group.

We both judged the effort to be a success. The committee itself was larger and more active than it had ever been, and we engaged in a number of new and potentially groundbreaking efforts, including the first attempt to introduce partner benefits at Ohio State. The benefits campaign effort failed, but it paved the way for later success. Beyond the committee's own activities, the co-chair structure put both of us on the Minority Issues Committee, and we were able to jointly elevate LGBTQ issues on its crowded agenda. But not everyone approved. Some of the GLB Committee members thought they would be a better fit than me for the co-chair spot and made it clear that they would like to displace me. But the bigger objection came from Robert.

During that year, Robert got a job in San Francisco. It was a long haul to move all the way to the West Coast, and being the dutiful best friend, I took some vacation time, drove all the way out with Robert, and spent a week helping him get settled. It was an intense week emotionally because of the transition in Robert's life and in our friendship. We knew we were going to live and operate in separate spheres, and it was only natural for our conversation to gravitate toward existential topics: who we were, what we were doing with our work and our lives, what was next for each of us.

One evening the topic of the GLB Committee and its agenda for the next year came up, and Robert revealed to me for the first time that he disapproved, strongly, of my being the co-chair of the group. The argument that followed was one of the most intense of my life. While some of his objections were not new to me, the intensity of his feelings was. That this was coming from my best friend and the person with whom I felt more aligned than any other was a real shock. That discussion changed me, changed our relationship, and sent me on a journey toward thinking much more deeply and very differently about ally leadership.

Allies in Leadership

Allies in leadership can produce critical challenges in social movements, practical and ethical dilemmas for both the ally and the movement organizations, and a host of emotional costs. And yet they can also work—helping to produce the outcomes the movement seeks. What are some of the aspects of leadership work in social movements that are altered by—and may even be incompatible with—ally leadership? What are the key dilemmas movements and allies should consider before moving an ally into a leadership role? When might an ally leader be successful in producing effective leadership? When shouldn't allies exert leadership? And why is this such a volatile issue?

Some of the challenges and dilemmas produced by allies in leadership are ramped-up extensions of dynamics produced when allies aren't in leadership, but merely when they enter the rank and file of a movement. I can't review these in detail in this short essay, but I've explored some of them elsewhere,[1] and they include ally disillusionment and disappointment at the reception of their efforts, lack of deep knowledge about the community they are attempting to serve, choices around exploiting and enacting privilege, ally ostracism from safe spaces, relative risks and costs, and paternalism.[2] Conflicts over these aspects of ally participation can threaten the continuing participation and contributions of both allies and beneficiary activists—and the very effectiveness of the organization. As fraught as those experiences can be, if an ally moves into a leadership role, not only are all these issues amplified, but new ones emerge as well. After all, leadership roles are, by definition, something quite different from rank-and file participation.

Later I'll explore what I have experienced as the five big clusters of enduring challenges for activist situations that involve allies-in-leadership. This year marks thirty years since I became involved in the Minority Issues Committee, and in those three decades of ally activism (across a huge range of environments, successes, failures, and relationships with all kinds of activists), these five leadership issues have repeatedly been the most salient— and pivotal to success. The experience that informs these thoughts is heavily conditioned on experiences with LGBTQ issues and groups, but they have also been ubiquitous concerns in other movements that employ allies, going back many decades.

Legitimacy and Identity

One of the paramount challenges any leader faces is establishing and maintaining legitimacy. A legitimate claim to authority is essential to producing effective action in all kinds of organizations.[3] Legitimacy in less rigidly hierarchical organizations (including and especially social movements and other voluntary associations) tends to rely heavily on endorsement from the membership—and that approval can be heavily contingent on identity markers. Can men, even demonstrably pro-feminist men, garner the support necessary to lead a feminist organization? Can white people have a credible presence leading a Black civil rights organization? Indeed, when I have asked lesbian and gay activists about leadership roles for straight people in their organizations, the typical response was "Are you kidding?" and "That would be bizarre!"

One reason allies struggle for legitimacy comes from the group's agenda: What does the membership think is its core purpose? If they believe that a primary objective, or even a singular aim, is to *build empowerment and identity*,[4] it becomes harder to fit allies into leadership. The fundamental notion of taking back the power for self-determination is both symbolically and substantially contradicted by putting someone from outside the community in a position of power. And really, this was Robert's biggest objections to my leadership of the GLB Committee. Empowerment and identity building was not only important for LGBTQ activist groups; he believed it was *the essential activity*—the core accomplishment that would enable everything else the movement wanted to achieve. That concept of identity and action was at the heart of Queer Nation's agenda as well: the movement needed people to come out, claim their identities as queer folks, and parlay that personal power into action on behalf of their rights and lives. It is extraordinarily difficult to imagine a straight white man, with many markers of privilege and advantage, standing up in front, exhorting, inspiring, or demanding that kind of action. It wouldn't be effective, and it wouldn't be right.

Instrumental Advantages

If a movement organization's goals are defined in more specific instrumental terms, however, allies *might* fit more comfortably into the leadership. If the aim, for example, is to produce legislation or policy, perhaps to introduce same-sex partner health benefits, the calculations can be quite different. Certainly, elite political allies can be of great assistance to movements,[5] and allies can be recruited because their networks will allow access to influential politicians, celebrities, or other power brokers. As I moved up higher in the administrative structures of universities over the course of my career, there was no question that I became a more useful ally because of my direct influence on policy changes and my access to others who could influence the outcomes, but also because of the symbolic value of having an open ally among the university's leadership.

Allies can also be particularly effective when it comes to externally facing leadership activities. For example, one key activity for the leadership of movement organizations is to acquire resources, the most important in social movements being people power and funding.[6] While beneficiary activists may be more successful raising money and soliciting participation from their own community, ally activists can be especially helpful recruiting and fundraising from other allies.[7] Indeed, when chasing resources and performing spokesperson functions, allies often get a credibility bonus, especially from other potential allies. For example, perceived similarity between communicator and audience produces trust in the communicator and the message.[8] Credibility is also enhanced when the communicator is making an argument that is apparently counter to his or her own interests.[9] So when an ally champions dismantling a privilege that she or he enjoys, the message is more likely to be convincing. And yet even when aware of these bonuses, the movement faces another dilemma: Does it want to exploit the privilege that produces these kinds of advantages?

Socioemotional and Task Leadership

Leadership roles are sometimes divided into two categories: those concerned with the social bonds and emotional well-being of the group's membership, and those concerned with the tasks and functions that the group must perform. With respect to the former, effective leaders often provide *inspiration* through their communication, both inside and outside the group. They need to provide motivation, energy, and confidence, both to their existing followers and to new people who may be willing to take up the cause.[10] Leaders also need to *encourage solidarity* among the membership so that they can support one another and act in concert. Internal solidarity is absolutely critical to producing the commitment to act in the face of difficult and costly circumstances—and that solidarity is conditioned on joint membership and identity. These kinds of socioemotional tasks can be a tall order for ally leaders, who lack the immediate, direct link to community and solidarity based on a shared sense of identity, history, and life experiences. Allies necessarily look in from outside the community and find it deeply challenging to produce bonds within a group in which they aren't organic members.

Leaders also need to *resolve conflicts* among members or between factions who differ on agendas and how to pursue them. Members of the beneficiary community, especially long-serving ones, often have considerable gravitas and authority they can wield in resolving disputes, but allies may not be seen as having the experience or insight to understand, represent, or adjudicate the issues at hand. Ally missteps in those moments are likely to reinforce the sense of otherness, as well as any notion that they should not be in leadership nor making decisions for the membership.

Consider one strategy that some have advocated—outing local celebrities—arguments about which have been volatile. Beyond the ethical dilemma represented by the practice itself, consider a situation where some of the LGBTQ-identified activists support the outing strategy and some do not, and it falls to the leadership to resolve the dispute and possibly even to make a decision for the group. Is it appropriate, in this circumstance, for an ally to make or even influence these decisions? And yet by taking on leadership, one may have committed to making decisions in contentious moments. Perhaps, then, if the ally is not situated properly to act in that moment, she or he never should have taken the mantle of leadership.

Leaders also have responsibilities associated with getting the work done. Task leadership, therefore, is an important part of the role, and the accompanying strategic decision-making can be fraught for allies as well. Consider a few of the strategic decisions that are commonplace in social movement organizations: leaders decide what is going happen and when—including implementing acts that incur substantial risks and costs; who to include in the membership of the organization and in the inner circle of key decision-makers; what the decision-making processes and standards will be; how the group might deploy privilege (by, for example, recruiting elite allies); negotiating the purpose of their collective activism (identity building versus influencing policy, for example); and deciding who or what is going to be the target of their actions.

These kinds of decisions can be challenging for any leader, but they can be especially so for ally leaders. Before deploying their troops, for example, social movement leaders must understand the situated lives of both their activist corps and the movement's intended beneficiaries. How will the proposed activities fit into the activists' lives? What is biographically possible? What kinds of risks and costs are tolerable for some people but not others? What is the level of commitment the leader can depend on? These factors all differ widely across the people involved, and a thorough appreciation of those profiles is essential to producing an effective and supported plan of action. Ally leaders are suspect on these grounds because they lack the dyed-in-the-wool life experience that can help provide that understanding and enable good strategic judgment.

Risks, Privilege, and Commitment

Leadership also entails taking on additional risks and costs. Leaders become the face of the movement and thus become targets for criticism, repression, countermovement activity, and even threats and assassination. And people are differently situated in their ability to endure these kinds of risks and costs. Often allies are ensconced in a buffer of privilege that makes their risks and costs less present and less damaging, making them better choices for some aspects of leadership. Straight allies, for example, suffer less targeting, fewer fears of losing their jobs, and so on, compared to an out gay man. Indeed, I argued to Robert that putting myself on the front line was a moral imperative. I felt compelled to step into the crossfire because that movement activity and what it would cost me was small suffering compared to the full-time lives of others.

At the same time, that risk profile can signal less commitment. Those who are putting everything on the line—their jobs, their family relationships, their bodies, and sometimes even their very lives—are thought to have unshakably deep devotion to the cause. They earn a special kind of trust and acceptance. Allies, however, are suspect on these grounds. Not only are they likely to incur less cost, but they also can more easily walk away from the costs and risks if they become too burdensome. Straight allies are never outed in ways that would damage them as a lesbian or a trans person, for example, and it is so much easier to go back into the ally closet. Having this problem in the rank and file is bad enough, but to be uncertain about the commitment and persistence of a leader is substantially more fraught.

Benefits

Just as leadership entails additional risks and costs to the leader, it also comes with benefits.[11] Potential rewards (especially if the movement is successful) include glory, prestige, visibility, and even fame—which might be parlayed into other benefits, including financial opportunities. Leaders might advance their careers as writers, intellectuals, politicians, or even well-paid activists. And of course, activism can produce the opportunity to rub elbows with the elite—politicians and celebrities who are aligned with the cause.

Whatever rewards might accrue to ally leaders, even meager ones, if they are allocated to allies, then they are not available to beneficiary activists. The ally has supplanted a member of the community. Indeed, this was a source of resentment some members of the GLB Committee felt toward me, even though the rewards were, indeed, small ones. Even when the rewards are unanticipated, they can be substantial. One student ally I worked closely with became a Rhodes Scholar after he graduated. He later reflected that his success as an ally activist had been pivotal in winning the Rhodes. Of course, it was not just his activism that had mattered, but it did provide a lift in his career—a lift that might have been even more important for someone with fewer privilege markers. Movements must decide, therefore, if they want to allocate their leadership rewards to relatively advantaged allies—and allies should reflect on the ethical ramifications of accruing these kinds of rewards.

In the end, promoting allies into leadership entails a number of calculations and considerations by both the movement and the potential allies. And it's got to be a lot more than just *here's a do-gooder with some extra energy*. Allies can be important, even critical, parts of social movements, both as rank-and-file participants and as leaders, but my experience suggests very strongly that they will be more successful additions if everyone goes into the relationships with eyes wide open. Thinking through some of these dilemmas and carefully considering the agenda and purpose of the group's activities will go a long way toward ensuring a match between roles and purposes for allies. What, for example, is the balance between instrumentally specific goals and identity-focused goals—in both the long and short term? And what sorts of compromises might need to be made in one set of objectives to increase the chances of success in another? Openly and broadly acknowledging these challenges, shortcomings, and compromises will not only increase the chances for productive ally engagement but will also keep the conversation active as the organization matures and its activity adjusts to a changing external environment.

Notes

1. Daniel J. Myers, "Ally Identity: The Politically Gay," in *Identity Work in Social Movements*, ed. Jo Reger, Daniel J. Myers, and Rachel Einwohner (Minneapolis: University of Minnesota Press, 2008), 167–187.
2. Although he does not use the term "ally activism," many of these dynamics are ably illustrated in Doug McAdam's account of the Freedom Summer drive, when northern white college students went south to help with voter registration drives in Mississippi. Doug McAdam, *Freedom Summer* (New York: Oxford University Press, 1988).

3. M. Zelditch Jr. and H. A. Walker, "Legitimacy and the Stability of Authority," *Advances in Group Processes* 1 (1984): 1–26.

4. Verta Taylor and Nancy Whittier, "Collective Identity in Social Movement Communities: Lesbian and Feminist Mobilization," in *Frontiers in Social Movement Theory*, ed. Aldon D. Morris and Carol McClurg Mueller (New Haven, CT: Yale University Press, 1992), 104–129.

5. J. Craig Jenkins and Charles Perrow, "Insurgency of the Powerless: Farm Worker Movements (1946–1972)," *American Sociological Review* 42, no. 2 (1977): 249–268. Sidney Tarrow, "States and Opportunities: The Political Structuring of Social Movements," in *Comparative Perspectives on Social Movements: Political Opportunities, Mobilizing Structures, and Cultural Framings*, ed. Douglas McAdam, John D. McCarthy, and Mayer Y. Zald (New York: Cambridge University Press, 1996), 41–61.

6. John D. McCarthy and Mayer N. Zald, *The Trend of Social Movements in America: Resource Mobilization and Professionalization* (Morristown, NJ: General Learning Press, 1973).

7. Sidney Tarrow, *Power in Movement: Social Movements, Collective Action, and Politics* (New York: Cambridge University Press, 1994).

8. Ellen Bersheid, "Opinion Change and Communicator-Communicatee Similarity and Dissimilarity," *Journal of Personality and Social Psychology* 4, no. 6 (1966): 670–680.

9. Alice H. Eagly, Wendy Wood, and Shelly Chaiken, "Causal Inferences about Communicators and Their Effect on Attitude Change," *Journal of Personality and Social Psychology* 36, no. 4 (1978): 424–435.

10. Doug McAdam, *Political Process and the Development of Black Insurgency, 1930–1970* (Chicago: University of Chicago Press, 1982).

11. Anthony Oberschall. *Social Conflict and Social Movements* (Englewood Cliffs, NJ: Prentice-Hall, 1973).

7

Organizing Dilemmas across U.S.-Based Social Justice Movement Spaces

alicia sanchez gill

The dilemmas for those of us organizing in these moments are too numerous to be contained in a single chapter. And for those of us who live at the intersections of many marginalized identities while working across movement spaces, our disappointments often feel like a deep well of lessons learned. During this time especially, of tremendous pain and transformation, we are seeing competing needs and demands; prison reform versus abolition, nonviolence versus radical defense, divestment from state institutions versus holding the state accountable to provide for *all* of our needs instead of the needs of the few, voter engagement versus voter fatigue and rightful distrust of electoral processes.

All of these tensions are occurring as our communities—that is, those of us who have been historically and contemporarily marginalized—are feeling the weight of the dual pandemics of anti-Black racism and a global health crisis. I write this as the deepest and most persistent inequities in the United States are laid bare *and* as we see incredible mobilization from communities under threat, the movement-building that will bring us out on the other side of crisis and closer to a more just world.

I write this as the COVID-19 pandemic and its aftermath bring to light the disparities in all of our systems—disparities that many have been fighting against long before the moments of shutdown and quarantines. Our systems are failing us because they were never designed to protect us, especially those of us at the margins. We are witnessing massive fissures in our economic foundation; the same workers now deemed essential in serving our communities have also been the workers fighting for a living wage. The people picking our food, stocking shelves, delivering mail, and keeping some semblance of normalcy during these extraordinary times, have also been unionizing, demanding safe working conditions, fighting for healthcare and other

alicia sanchez gill, *Organizing Dilemmas across U.S.-Based Social Justice Movement Spaces* In: *Wicked Problems*. Edited by: Austin Choi-Fitzpatrick, Douglas Irvin-Erickson, and Ernesto Verdeja, Oxford University Press.
© Oxford University Press 2022. DOI: 10.1093/oso/9780197632819.003.0008

basic needs. The people most likely to be impacted by COVID-19 deaths are also least likely to have health insurance, most likely to be making low wages or working in essential roles, and were already experiencing health disparities due to segregation, racism in medical institutions, environmental degradation, housing instability, and poverty. Yet, in response to the pandemic, we also witnessed people being freed from jail and detention; real movement toward affordable child care, universal sick leave, healthcare for all, living wages for child care workers; eviction, utility, and foreclosure moratoriums; universal emergency shelter; increased resources to Indian Health Service; and more.

In addition to the devastating effects of COVID-19, we watched as millions of people poured into the streets to demand justice for George Floyd and a complete end to the murderous, white supremacist systems that killed him—the same systems that killed Tony McDade, Ahmaud Arbery, Breonna Taylor, and many, many more. The demand to defund police has placed words like "carceral" and "abolition" into the common lexicon, in a groundswell not seen in a generation. The work of this moment is building on longstanding movement work by organizers, particularly Black organizers and organizers of color, and the continued leadership of these communities is paramount to justice.

As conversations about what is needed in this moment swirled, we saw immediate fractures in movement spaces—particularly around what to do about the question of policing. Just a week after the uprisings began, a group of activists put out a call to action called "8 Can't Wait." The campaign points to research that suggests police departments' adoption of eight measures would reduce police violence by up to 72%. Less than a week later, abolitionist activists created a website, *8 to Abolition*, to refute the goals of police reform and to instead move communities closer to abolition by divesting in police departments.

Core to values of abolition is the belief that policing, like the military, is a system founded on racial violence, misogyny, and power imbalances that cannot be either redeemed or reformed. 8 Can't Wait has since deemed their campaign a *harm-reduction* approach to racist policing while using tactics and reforms that have already been widely implemented to little improvement. But what organizers on the ground, the people closest to the harm, are calling for is abolition, not a piecemeal or reformist approach to ending state violence. Organizers on the ground are clear that empty platitudes and streets painted with the words "Black Lives Matter" mean nothing when Black people are being murdered for simply existing.

Abolitionists are using the momentum of this moment to fight for long-standing demands, diverting resources from policing and into community services, safe and affordable housing, healthcare, education, community-based violence interrupters, and culturally appropriate social services disconnected from the state. Abolition calls for more than a 72% reduction in state violence; instead abolition demands an end to the carceral violence of policing, prisons, the Immigration and Customs Enforcement deportation machine, and war. This dilemma comes up again and again, from voting to partial criminalization versus full decriminalization of sex work. It is often the people with more access to resources, visibility, and funding who dominate policy reforms and the attention of media. However, this moment is calling on us to be bold and to settle for nothing less than drastic shifts to the status quo.

From the time I began organizing with INCITE!, a network of radical feminists of color in the United States and Canada organizing to end state violence and violence in our homes and communities, through my role today, organizing philanthropy and people with wealth to move resources to Black, Indigenous, and people of color organizers, our challenges have been numerous. In this chapter, we explore several stories that highlight the hurdles to organizing across movements. The organizer/writers I have collaborated with more often raise questions than provide answers. The reason for this is that the work is messy, and we are constantly finding our way toward our North Star, and toward one another. We are modeling what it means to bring others along and practice shared storytelling. While the stories vary, there are consistent themes around what it means to organize as a person of color in a white supremacist world—even in our social justice spaces. These stories discuss power, colonization, anti-Blackness, and misogyny.

My own story is a lot like theirs. I came to the work of organizing as a survivor, looking to create spaces of survivor safety and ultimately to shift the conditions that allow for violence to happen, without relying on carceral systems, knowing that those systems have been antagonistic *at best* toward many of the communities of which I am a part. I have played multiple roles within the organizing ecosystem, starting in direct services (counseling, case management, conducting support groups); doing direct actions, community organizing, and base building; policymaking and advocating with lawmakers on the local and national level; research; and now moving resources to organizers through my role in philanthropy.

In every one of these spaces, the issue of funding arises. It is often funding, rather than movement leaders, that gives the work its particular direction. Whether it's small grassroots groups in a coalition fighting over the scarcest of resources from foundations, nonprofits being willing to depoliticize their work for government grants, or incredible base-building and Black-led movement work being chronically underfunded, we know that funding can be an asset or a detriment to radical work. Funding may determine who is worthier of funding based on outdated and biased beliefs about trust and respectability. It's no surprise, then, that organizations led by people of color, especially those that engage in systems transformation work, are underfunded compared to white organizations. Funding also decides what issues are "hot" right now. As I write, what's suddenly hot is COVID-19 response, support for Black-led organizing, and dabbling in abolition, leaving organizers and organizations scrambling to fit within these new criteria. Funding may also lead to a desire to "professionalize" movement spaces, valuing lawyers, licensed social workers, and policy professionals over those with lived experience who may not have degrees or formal credentials. How do those of us with the very noble intention of reallocating resources from the wealthy to our communities via philanthropy do that in an ethical way that does not reinforce gatekeeping, hierarchy, and power imbalances, and what is the role of philanthropy in movement building?

My belief is that it is philanthropy's role to move resources, trust the leadership of the people it's meant to benefit, and get out of the way. That may look like:

- Ensuring grantmaking decisions come from a community advisory board rather than funders/people with wealth.
- Reducing the reporting requirements and accept reporting that doesn't prioritize the written word over other mediums (videos, on-site appearances, art, phone calls).
- Investing in multiyear grants.
- Putting people impacted by the issues in leadership.
- Minimizing application processes (and questioning the intent behind your distrust of organizers of color).
- Committing to investing as many resources as possible in moments of crisis.

This list is by no means exhaustive. The dilemmas of this moment will not be solved through philanthropy alone, but philanthropy has a role in not obstructing movement building and in challenging the status quo. What I have learned in my short time in philanthropy is that our work must rise to meet these moments of change in solidarity and partnership. And we must follow the lead of the organizers on the ground as they move policy through direct action, base building, and advocacy.

There is inherent tension while working within a system you don't believe should exist at all—a sector that would not exist in this iteration without extreme wealth hoarding and lack of safety nets for so many. The dilemma of whether to remain within the system and help to mitigate the harm it has caused or to work outside of these systems altogether continues to offer both tensions as well as an invitation for self-reflection. I have decided to do what I can where I am to disrupt the violence of philanthropy. As a young Afrolatina, queer executive director of a social justice fund, I believe that my role in movement spaces in this time is to redistribute wealth to the organizers on the front lines who need it most, right now. As I grapple with these questions in my corner of the work, organizers are also facing dilemmas. Here are other stories from organizers today.

Lessons Learned from Organizing in Leftist
Spaces as a Muslim Woman
Darakshan Raja

One of the biggest challenges in organizing as a Muslim woman in leftist spaces is that while some have a theoretical understanding of the state's repression toward organizers and activists, there is more work to be done to truly create spaces that understand how state violence, gendered violence, harassment, and the surveillance state target Black and Brown Muslim women and femmes.

One example I want to share is from a former organization I worked in. I expressed to my board that as a Muslim woman, in the Trump era, protesting in the streets meant I had to think about the ways the surveillance state would retaliate and target me individually given my immigrant Muslim identity, as well as the organizations I had previously been affiliated with. I still recall having a white woman on the board state that this organization, a white organization in the peace and justice movement, had survived being targeted under the FBI's infamous counterintelligence program, COINTELPRO. Note that she was applying the experiences of white leftist peace and justice, antiwar activists to me as a Muslim woman. Unfortunately, the board of my organization didn't see my concern as valid. On the contrary, they began portraying me as paranoid and as centering my own experience over others. Eventually they communicated my fear to external community organizers, representing it as my excuse for not being more present to support protest.

I share this story because I believe our movement spaces have far more work to do in actually hearing the valid threats and safety concerns Black and Brown Muslim women and femmes raise around what we need in order to be in movement spaces. I know that my experience isn't unique and that our movements have become complacent in using our labor, thought leadership, visions, and work to highlight the intersectionality and inclusivity of movements but have yet to do the hard work of building structures that can actually hold us and offer us the safety, care, and security we need to be present and agitate. Not everyone's dissent is treated the same way, and it is wise to be thoughtful about what infrastructure is needed to protect, support, and really hold us in movement spaces.

Families for Freedom
Janis Rosheuvel and Donald Anthonyson

Families for Freedom (FFF) was founded to resist the brutality of the mass deportation system that emerged following passage of the anti-immigration laws that legalized arbitrary detention, fast-track deportation, and family separation, known as the "1996 Laws,"[1] and the post-9/11 criminalization of migrant life. FFF was organized in 2003 by and for people facing and fighting deportation. Members are Black, Asian, Latinx, undocumented, documented, with and without criminal convictions. Our members maintain the mature political analysis that no one should be deported, regardless of status or past criminal convictions.

This vision is based on the principles of family unity, second chances, and transformative justice and with an understanding of the systemic ways immigrants are criminalized. In the mid-2000s, FFF members wrote the Child Citizen Protection Act to enable immigration judges to consider whether deporting someone would harm their U.S. citizen children. This bill had an unwavering political vision given the threats deportation presented to our members, their families and communities. But we were often asked to compromise during this time. The questions were framed like this: *Shouldn't people with really bad convictions be deported? Maybe only green card holders shouldn't get deported? Aren't undocumented people without criminal convictions more deserving?* In coalitions, op-eds, and organizing meetings and at policy tables, our members responded resoundingly: Stop deportations *now!* Period. Their work affirmed that no human being is illegal but also that families should not be separated, regardless of status or conviction.

The firmness of this position put FFF in opposition to many in the mainstream immigrant rights movement. Many thought we as a movement had to agree to more enforcement against our people (detention, deportation, border militarization, etc.) to see even the most meager immigration reforms in our lifetime. This compromise did not work. The current state of the immigration system leaves us bereft. The enforcement machine that took hold in the mid-1990s has metastasized. It is also clear that holding an unwavering line on stopping detention and deportation, as FFF members and other like-minded allies did in the mid-2000s, led us to today's renewed campaigns to fix the 1996 Laws, shut down detention centers, stop caging children and families, protect legislation like the Deferred Action for Childhood Arrivals and Temporary Protected Status, and stop the deportation beast in its tracks. *Often, not compromising is the only way to maintain the fundamental vision that families belong together, not in cages and not separated by deportation and exile.*

A Grandmother Was Killed in Atlanta
Zuri C. Tau

A grandmother was killed in Atlanta. An old woman who, living her last two decades in one of the oldest and poorest communities in the city, had undoubtedly seen too much. Yet she was probably a lot like your grandmother. I imagine Ms. Johnston had a very neat kitchen, with pictures of her grandchildren placed at regular intervals on her walls and long trailing pothos plants seeking out sunlight through heavy curtains. One night, Atlanta police burst into her home and opened fire.

Three years later, in 2009, a small group of multigenerational Black and Latinx women, men, and gender-expansive folks started organizing against police brutality and for alternative mechanisms for community safety. At the time, we didn't realize how unlikely our alliance was—it just felt right. I had never been a community organizer, but soon power analyses, press conferences, and speaking to city council became second nature. We were galvanized by the creation of a citizen review board that was formed to provide police oversight in response to Ms. Johnston's murder. Our group, Building Local Organizing for Community Safety (BLOCS), was comprised of poor folks from Vine City and Old 4th Ward, and middle-class twenty-somethings like myself. The establishment of the citizen review board made headlines, yet we soon discovered that police were refusing to be held accountable—they skipped out on hearings or refused to speak when they showed up. Those of us with full-time jobs could take time off in order to show up to the midday hearing, but those who had the most to lose were often working two or three jobs to survive. A few years into our work, one of our disabled members was assaulted by the police in front of their apartment. The lawyers in our group rallied, and the city paid damages. We eventually won subpoena power for the review board; after so much work we finally felt we had made a difference. The next voting cycle came, and we were devastated when the city council revoked the review board's power. Politics was bitter; I learned the hard way. *Was organizing worth it? For us? For those who were most vulnerable to abuse?* I never asked the members who lived in Ms. Johnston's neighborhood what they thought. I left BLOCS soon after to organize diversion programs for first-time offenders. I still wonder if I gave up too soon or if our work meant less because those who were most vulnerable were not the most visible. *Were we the right organizers for the fight? Was an impotent review board better than no response at all?*

What Is Integrity to a Traumatized Person? Or,
How to Hold a Trauma Survivor Accountable
Latishia James-Portis

I thought my understanding of the nervous system was pretty good until a few months ago. I knew about flight, fight, freeze, and fawn. I knew that it did not matter whether or not a perceived threat was real; it could trigger our flight, fight, freeze, or fawn responses just the same. I knew that once a person is triggered, they could not be expected to perform higher thought functions. I knew they would likely be reduced to their basal behavior of survival or defense. I also knew that even triggered people, whose instinct to survive could result in their behaving in ways that caused harm to others, should be held accountable for their actions.

I thought I knew all of this. And then one day all of my training and all those years of working with trauma survivors could not be accessed. It couldn't be accessed because I myself was in a full body trauma response. It caught me totally by surprise and activated me in ways I didn't even know were possible. And just like that, I became the hurt person hurting other people.

How do you hold yourself with integrity when you do not know who your self is? When you do not want to be reduced to your trauma, but when your trauma is very actively keeping you from showing up as the person you claim to be?

Aligning values and behavior—what we believe and how we act—is a dilemma. It is hard to admit when you fall short of behaving in alignment with your values. Especially when your life's work is committed to the healing, restoration, joy, pleasure, and liberation of some of the most marginalized members of society. Life as a trauma survivor in essence means being in a near constant state of assessing whether you are engaging with the world in the same way others are engaging with the world. And unfortunately, most of the time the answer is no. For the trauma survivor there is frequent scanning of one's surroundings for threats, and our bodies and minds are ready to spring into protective action at even the slightest hint of danger. Whether the threat is real or perceived, we're ready to fight, flight, freeze, or fawn.

For trauma survivors, important work is needed as we learn to cope, to be in the world, and to cultivate a sense of safety. It takes work to not see those who support us as threats to our survival. But telling the difference is not always easy. I identify myself as a healer, an advocate, and a facilitator in the movement. I have worked in gender and reproductive justice since I was a teenager and have always taken a survivor-centered approach in my work. I never thought that I made *excuses* for survivors who engaged in harm—until the survivor hurting others was myself. If someone says that you have harmed them, even if you disagree, but if you truly operate from a trauma-informed and healing-centered position, there is value in accepting the experience of one who says you have harmed them. Even if you disagree you will accept what they say at face value, apologize, and work to make sure you do not enact that same harm again in the future. But what happens when the harm occurs while you yourself are in a triggered state? What if you can't remember the harm? What if you genuinely believe that if it were not for your own trauma response the harm would never have occurred?

I now feel immense remorse for the ways I behaved. And that is the point. I, triggered or not, behaved in those ways and must therefore be held accountable. As an abolitionist committed to healing and survivor-centered solutions to harm, I hope more movements can embrace the nuances of conflict and harm and create environments where we do not treat things as though they are cut and dry. While I hope I never again act as I did in this situation, I cannot give that guarantee. We are all capable of enacting harm. We will all experience harm. May we all be humble and gracious enough to remember this when the time comes.

Movement Masculinity
Travis Akil Brookes

Masculinity, masculinity, masculinity. Shit's toxic. I can read, I can go to the workshops, I can have the conversations, but if I'm not giving myself the grace to unlearn masculinity then I'm more of a problem than a "solution." For three years I took part in a program aiming to get participants to rethink their masculinity. I first participated, then facilitated, and eventually ran the program. But while I worked with folks to undo their harm, I forgot to undo my own harm. I took up this martyrdom ideology, thinking I needed to sacrifice myself to undo masculinity, believing that if I kept showing up, that would be the work. Those ideologies are patriarchal as hell. It was performative. It was self-serving. It was causing more harm to me and others. It wasn't sustainable, and eventually I broke down.

In the midst of that moment, in the middle of that lost feeling, is when all of the self-work kicked in. Folks had already put in the labor, folks were already actively supporting me, folks were actively teaching me, and I was slowly learning how to build my communities of support, slowly trying to address violence, slowly trying to practice enthusiastic consent, boundaries, and vulnerability. This wasn't some epiphany or "aha" moment. It was a realization that I wasn't the solution. It was the labor of Black, trans, and queer folk that helped me to see that where I am now, where I was then, and where I will be is all a part of the journey. And if I'm really about unlearning this shit, then I gotta recognize that I'm going to fall down, I'm going to fail, that I need to center healing, and that I should not do any of this alone—independence is a deeply entrenched belief of masculinity and white supremacy. The work is messy, the work is important, the work is hard, and the work was being done before I even took a workshop.

Organizer in Washington, D.C.

Nicole Newman

The idea of wanting to hear from the community in times like this is very exhausting. White folks in organizing and advocacy spaces suddenly see the validity of lived experience and then want to hear from people impacted, though many Black people have been restating the same needs, root causes, and interventions for years. In this moment in particular, the emotional labor White folks are asking me to do, by opening up my relationships to them and then the translations of our experience that they require, is transactional and performative.

For outsiders, there is a constant flattening of our experiences and a belief that there is one way to be Black and working class. As someone who now has educational privilege it places me as the central point, as the gatekeeper, and often asks me to make sweeping generalizations for a reality I no longer live and am not always best suited to discuss.

It is both intellectually lazy and disingenuous *but safer* for me to speak on behalf of folks "who are from here" or "who live east of the river" than it is for white folks to examine their bias and why they must always understand before they feel or dissect problems rather than working in their community to prevent harm. I feel constantly on display, like if my experiences don't match up with what it means to be working class or Black in D.C. then I am either dismissed or asked to validate those experiences to see if they are "widely felt."

Conclusion

These stories highlight some of the dilemmas that organizers of color grapple with: how to embody masculinity in a way that does not harm, how to speak for our communities without being tokenized, how to organize on the front lines while being protected, what happens when our behavior is out of alignment with our values, and when to step back and let others speak for themselves. And here is the dilemma: how to show up, in the midst of debates between reformist approaches and real mobilizing to push harmful systems toward a more radical vision of the world in the midst of a pandemic, in the midst of highly visible acts of state violence against Black lives.

I am resolved that this is not the time to play small with our demands. We have to believe a radically different world is possible and work to create it, even if we haven't seen it yet. There is no peacebuilding without justice; there is no conflict resolution in the absence of restoring what has been taken through colonization, enslavement, and violence; and there is no movement that does not bring others along. So we struggle together, we share our stories, and we dream of new worlds even in our apocalyptic reality. That is our work.

Note

1. These are the Antiterrorism and Effective Death Penalty Act, and the Illegal Immigration Reform and Immigrant Responsibility Act.

8

The Ones Who Walk Away to Stay and Fight

Philip Gamaghelyan

In *The Ones Who Walk Away from Omelas,* Ursula Le Guin depicts a society of near-perfection where the harmony and happiness of all comes with a price, one that is paid by a child who is confined to a dungeon and perpetually tortured.[1]

Omelas is less a dystopian fantasy than it is a mirror reflecting the actual face of capitalism. Those of us who enjoy relatively comfortable lives know, as do the residents of Omelas, that our bliss rests on the shoulders of children who labor and suffer in developing countries, as well as those within the less privileged neighborhoods of our own country, and that poverty, sexism, racism, climate change, and other forms of structural violence are systems designed to benefit the privileged few.

In this fiction piece, as in real life, most people choose to conform and accept the price paid by the child. The few who do not, "walk away" to a place unknown. Le Guin, acclaimed by some, has been criticized by others for painting a picture of a bleak reality, only to "walk away" without addressing the problem.

Nora Jemisin's *The Ones Who Stay and Fight* serves as the most notable rebuke of Le Guin's solution.[2] Jemisin advocates for a path chosen by many committed to social justice. Yet a dilemma persists: Is it possible to both stay in Omelas *and* fight? In other words, can one simultaneously be part of the system, benefit from it, and still effectively work for meaningful change? Transposing this question to peacebuilding: Can scholars and practitioners work within a system that perpetuates violence and avoid co-optation? Can they perform the ethical balancing act of maintaining integrity while benefiting from funding streams with "strings attached"? Le Guin leaves us with a significant puzzle: Is the walk away from Omelas a *flight* or a risky (yet ethical) path toward the *fight*?

Philip Gamaghelyan, *The Ones Who Walk Away to Stay and Fight* In: *Wicked Problems.* Edited by: Austin Choi-Fitzpatrick, Douglas Irvin-Erickson, and Ernesto Verdeja, Oxford University Press. © Oxford University Press 2022.
DOI: 10.1093/oso/9780197632819.003.0009

The Ones Who Stay and Fight

Allow me to examine Jemisin's proposition and the dilemmas of those who stay and fight in the context of the conflict-ridden region of the South Caucasus. The ones who work from within the system are not a monolithic group. Some of them operate on the (neo)liberal end of the spectrum, including employees of the World Bank, the United Nations, development agencies, foundations, and large nongovernmental organizations (NGOs). While they acknowledge some flaws, those who work within the (neo)liberal paradigm generally have faith in the current systems and work to sustain and improve them.

Others have chosen to "save trees while the forest burns." The series of interviews I conducted for this chapter, as well as my previous research,[3] revealed a surprisingly large number of peacebuilders who are fully aware of the structural violence inherent in the existing systems but work to maintain these systems in the interests of financial stability and the growth of their organizations. At the extreme, this category includes a cynical few who see perpetual violence as a guarantee of their own continuous employment. Finally, and of particular interest to this chapter, are those in the third category, who see the violence inherent in the current systems and stay, to fight and transform them from within.

There are blurred distinctions between these categories. Anyone who "stays in Omelas," even with the purest of intentions, is co-opted by virtue of benefiting from the suffering of the proverbial child. And even the most hardened opportunists among peacebuilders often "do good" by way of their job description. What distinguishes practitioners within the third category from the first two, therefore, is not their impact but their intention. These individuals work with and benefit from the existing systems yet persevere with the intention of transforming them. Implicitly or explicitly they reject the dichotomy between structures and agency, considering structures to be central to the construction of violence, yet viewing themselves as agents capable of reshaping and transforming them.[4]

This path is fraught with ethical dilemmas. Philosopher Ivan Illich criticizes the fields that professionalize interventions concerning vital aspects of human life as crippling institutions that curtail people's ability to provide for their own needs and that turn societies into passive clients. Professionals define both the norm and its deviations while simultaneously holding a monopoly over the presumed remedy, creating localized problems that they then "solve" while leaving the structures of violence intact.[5] Illich writes about education and health. The professionalized peacebuilding field that solves local problems but fails to address larger systems of violence also fits the bill.

I am not proposing to de-professionalize the peace sector. I read Illich's critique as a warning rather than a verdict. Both direct and structural violence are well institutionalized and legitimized in academia and policy-making.[6] Processes of war-making stand on strong infrastructure, feeding on the bottomless well of defense funding, their roots deeply entrenched in philosophical ground ranging from the Hobbesian war of all against all to the contemporary *offensive realism*.[7] It is impossible to counter the breadth of war structures with the enthusiasm of motivated activists alone. Professionalized war-making must be countered with professional peacebuilding that works in lockstep with local activists and those who are affected and marginalized.

The challenge, however, lies in preventing the co-optation of peacebuilding and its reduction into a mere rhetorical device and a cover for war-making. Anthropologist Laura Nader has been a vocal critic of the early iterations of the conflict resolution field, particularly the now mainstream Alternative Dispute Resolution (ADR) approach, for its promotion of "harmony ideology." Nader asserts that ADR pacifies those subject to structural violence and functions to maintain an unjust status quo.[8] Prior to Nader, anticolonial political philosophers Jean-Paul Sartre and Frantz Fanon were ruthless in their condemnation of nonviolence as a strategy expected exclusively from those who are colonized and violated, a position reiterated in this volume by Tony Gaskew.[9] The more recent disputes between conflict resolution and human rights fields saw the latter sounding similar alarms.[10]

The challenge for peacebuilding is structural: the resources that currently sustain the field are located within the structures that function to uphold hegemonies and produce profit, such as the nation-states of the Global North and multinational corporations. Peacebuilders who put profit and competition ahead of peace, therefore, are not outliers in this field. The unethical and outright corrupt behavior that promotes "harmony ideology" while maintaining violent structures is rewarded and incentivized, raising a key dilemma: working outside the existing structures tends to be impractical, yet working within them opens the door to co-optation. Can we professionalize but not be co-opted? This chapter explores this question, wrestling with it in a particular place and time.

The Ones in the South Caucasus Who Stay and Fight

Few become peacebuilders in order to enrich themselves, for in conflict zones where the dominant discourses are nationalistic, peacebuilding is dangerous and not particularly profitable. Most begin this work with a genuine motivation to prevent violence and enact change.

I recently convened two focus groups: one consisting of current and the other of former participants of cross-border dialogue projects in the South Caucasus.[11] Between these two groups, I found the differences in attitudes regarding the peace and conflict field striking. Most of those currently involved in dialogue were idealistic, eager to promote peace, and assumed that peacebuilding organizations and individuals "practiced what they preached" and prioritized ethics and integrity above profit and organizational interests. The focus group of former dialogue participants, however, revealed a more cynical attitude. Many of them had worked for peace but grew disillusioned with the unethical practices they witnessed. A longitudinal study of Israeli and Palestinian dialogue participants presented similar conclusions.[12] What were the reasons for the disillusionment?

The region that descended into ethnically framed violence at the same time as the liberal world order delivered a knockout punch to its alternatives and proclaimed the "end of history,"[13] unsurprisingly, received the "kitchen sink" of liberal peacebuilding. A number of international NGOs (INGOs) headquartered in Western European capitals entered the South Caucasus in the early 1990s with a mandate to build civil society, promote human rights, and help develop liberal-democratic institutions. They prioritized close coordination with local authorities, seen as both the direct recipients of the know-how of the INGOs on the road to liberal-democracy and as key stakeholders whose approval was essential to building local civil societies.

Yet by the 2010s, hopes of democratic development in the South Caucasus began to dwindle. Azerbaijan and Armenia devolved into authoritarian states, and Georgia struggled to establish a stable democracy. The governments ignited their war propaganda machines and showed little appetite for a genuine peace process. They acted to marginalize or suppress independent civil society and formed GONGOs—government-controlled organizations formally registered as independent NGOs—to usurp the peace process.[14]

Despite these authoritarian developments and human rights abuses, including those levied against the local partners of the INGOs, the latter continued to prioritize relations with local governments. This relationship had evolved, revealing the quagmire of an unspoken requirement of clearing strategies and participants' names with the local authorities, resulting in the creation of informal blacklists of "inconvenient participants" and the tacit promise of a governmental veto on any transformative work.

Close coordination with the local authorities coupled with the presence of a relatively large number of INGOs in this relatively small region proved detrimental to the development of a viable local peacebuilding infrastructure. As a diverse civil society began to emerge in the late 1990s, the INGOs and a handful of local NGOs formed consortia, monopolizing the funding streams and leaving the subcontracts from the consortia as the only readily available resources for most local activists and smaller NGOs. This monopolization has been particularly pronounced in the contexts of Turkey-Armenia and Nagorno-Karabakh conflicts, where by the 2010s the diversity of the 1990s funding streams was replaced with large European Union grants administered through two consortia: the EU's Support for the Armenia-Turkey Normalization Process (ATNP) and the European Partnership for the Peaceful Settlement of the Conflict over Nagorno-Karabakh (EPNK), respectively.[15] From 2010 on, the consortia processed millions in funding, presenting a threefold challenge for the development of local peacebuilding.

First, diversity in funding was diminished. With the EU allocating millions to peacebuilding, other donors who had previously worked in the South Caucasus deemed the field financially saturated and withdrew.

Next, the consortia's monopolization of EU funding streams and the parallel withdrawal of other donors created the perception among the smaller NGOs and activist groups that the consortia subcontracts were their only lifeline. This perception resulted in solidifying the control wielded by the consortia over the smaller organizations and activist groups in exchange for a symbolic percentage of their overall funding. While each funding stream for the consortia included built-in small grant schemes, these constituted a meager part of the overall funding, ranging from 1% to 5% of the total pool.

In other words, year after year 95% to 99% of the total funds, amounting to millions of euros, remained in the hands of a handful of consortia members augmenting their power. Meanwhile, a ceiling of 20,000 euros (per project) was established for those lucky local actors who qualified for subgrants.[16] This great disparity between the consortia members controlling the funds and the smaller organizations pushed most local NGOs and activists to the margins or out of peacebuilding entirely.

Finally, the monopolization of peacebuilding in the South Caucasus at the hands of the consortia had not only financial but also political ramifications. Throughout the 1990s and early 2000s, the EU struggled to gain a foothold in the official peacebuilding track, one dominated by the Russian and U.S. governments. The EU's role was limited, supporting unofficial processes and initially working with little coordination with the local governments. The advantage of this approach was the relative independence of EU-funded efforts and strictly voluntary cooperation with local governments. The August 2008 war over South Ossetia involving the Georgian and Russian governments would bring a change to this dynamic. The EU, having acted as a mediator between Russia and Georgia, appointed a Special Representative for the crisis in Georgia. In 2014, the position was renamed as the EU Special Representative for the South Caucasus and the crisis in Georgia, with the purview of the role expanding to concern Nagorno-Karabakh as well. As representatives of an intergovernmental body, the special representatives prioritized their relations with the local authorities. Subsequently, to continue their work, the peacebuilding organizations funded by the EU were to attest, directly or indirectly, that their personnel and activities were acceptable, and better yet loyal, to the ruling regimes. Effectively, the civil society efforts saw their independence sacrificed to the EU's increased role on the official track.

The interconnected structural challenges of (a) decreased diversity of funding sources, (b) the monopolization of the existing funds at the hands of the consortia and the marginalization or exclusion of smaller NGOs and activist groups, and (c) the right of near-veto over civil society efforts granted to local authorities by the special representative resulted in the hollowing out of the peacebuilding field by the early 2010s. Many peacebuilders were either pushed out or walked away. Those who remained commanded millions in funding, implementing projects that remained in the good graces of and closely monitored by local, mostly authoritarian governments.

The challenges from the consortia to the survival of the local NGOs and activist groups extend beyond the structural level. On the individual level, many of the INGO leaders involved in peacebuilding in the South Caucasus adhered to ethical practices and prioritized the needs and interests of their local partners. Others, however, engaged in predatory practices aimed at sabotaging the work of local NGOs. Practices of suffocation included lobbying the donors not to fund local NGOs directly and to instead provide subgrants through the consortia; establishing control over growth of the small NGOs through limits placed on their activities; hiring the most resourceful members of emergent NGOs and preventing the latter from gaining traction; offering minimal subgrants to emerging organizations and restricting their growth through excessive reporting requirements; putting limitations on the grantee's allowed program activities to avoid "competition" with the consortia members; and others. Further, many independent voices were excluded and potentially transformative activities marginalized through veto imposed by local authorities directly or via consortia, ensuring peacebuilding efforts' conformity with system maintenance.

The monopolization of peacebuilding funding by the consortia and their close coordination with authoritarian governments presented aspiring peacebuilders with a stark dilemma: to stay but not fight, accepting subgrants or jobs at the consortia, effectively conforming and contributing to system maintenance, or to walk away to a place with no known prospects for employment or funding and with no charted path to fight for change.

The consortia model has been criticized from outside the field by journalists[17] and watchdogs[18] with limited impact. Critics have pointed to the model's inefficiency, the monopolization of peacebuilding by a handful of actors, and the near-absence of meaningful engagement of local actors. The local peacebuilding community itself, however, long stayed silent, in part because of local peacebuilders' perceived financial dependence on the consortia, and out of concern that public criticism from within the field would play into the hands of nationalist groups and others within the war structures. Additionally, mechanisms and mediums for raising concerns and advancing change remained elusive.

In 2018, a group of ten emerging authors from Armenia, Azerbaijan, Georgia, Russia, and Turkey proposed to the editorial team of the *Caucasus Edition*, an analytic journal where I served as an editor, to publish a series of articles critiquing the state of the field. They were all peacebuilders who had attempted to stay and fight but grew increasingly frustrated with the unethical practices within and constraints imposed by the existing system and decided that it was time to walk away. Influenced by the proliferating academic critique of liberal peace, as well as by postcolonial and feminist theory, they were ready to break ranks and articulate their concerns publicly.

Initially the editorial team hesitated. Our concerns mirrored those I have outlined; we were worried that criticism from within could play into the hands of nationalists and weaken an already dwindling peacebuilding field. Further, because the parent organization of the *Caucasus Edition*, the Imagine Center for Conflict Transformation (Imagine Center), where I serve as a board member,[19] was itself a midsize peacebuilding actor in the South Caucasus and a beneficiary of many existing grant schemes and was on a path of joining one of the EU-funded consortia, our donors and partner organizations could perceive the publication of critical articles as a politically or personally motivated attack. Finally, our past experience indicated that public criticism of the major donors and consortia members might result in attempts to politically marginalize and financially hurt both the authors and the editors of the publication. The dilemma of whether to publish or to reject the proposed series of articles was thus as much political and ethical as editorial and academic.

Following deliberations with the prospective authors, the editorial team concluded that moving forward with the articles was the politically expedient and ethical choice. As the authors intended to criticize the Imagine Center itself, along with other NGOs, for its postcolonial and exclusionary practices, preventing the publication of such articles on political grounds would amount to an act of censorship. More important, the authors themselves were local peacebuilders committed to the transformation of the conflicts and structures of violence in the South Caucasus. Most of them had collaborated with or worked in the organizations and donors they were proposing to critique. Their motivation was neither the defamation nor the dismantlement of the field, but its rethinking, reconceptualization, and reconstruction.

The ten authors who had chosen to walk away formed three groups, each targeting a related but distinct area of the field. One group focused on large funding schemes and the consortia of INGOs that had monopolized the field. These co-authors advanced the postcolonial argument that current funding structures encourage the "enlightened" organizations of the Global North to advance the liberal-democratic agenda on "uncivilized others" through the control of the local civil society. They further argued that the international consortia used the rhetoric of "turn to local" and "bottom-up" approaches to obscure their "civilizatory" missions[20] and top-down practices that treated the locals as objects of intervention and not as stakeholders with their own voice and capacity.[21] The second group of co-authors critiqued the implicit reproduction of patriarchal structures by women's rights NGOs.[22] The third examined the practices of local gatekeepers, the big and medium-size local peacebuilding and women's rights organizations who had joined the international consortia and formally represented the "local voice," while functioning to keep resources out of the hands of smaller NGOs and activists.

The co-authors critiqued the rigid administrative criteria and foreign-language applications distributed by these organizations in calls for subgrants that made the majority of grassroots activists and small NGOs ineligible for funding. In line with Illich, they argued that the bureaucratization, or "NGO-ization," of activism had two repercussions: it shifted the agency away from grassroots activism and toward professionals employed by the NGOs, while simultaneously forcing these few overworked NGO employees away from grassroots activism and toward administrative-technical office tasks.[23]

The power imbalance between NGOs based in the Global North and the activists in the Global South is not a new phenomenon, nor is it confined to the South Caucasus. Already in 1978 Jeffrey Pfeffer and Gerald Salancik had proposed resource-dependence theory, thus naming the challenge.[24] Over the next few decades, numerous studies examined the problem from various angles. Seth Cohen pointed to the detrimental influence that resource dependency and the resulting competition for scarce resources have on subverting the mission and values of the southern NGOs,[25] and Michelle Shumate and Lori Dewitt examined the impact of the social media revolution and came to the conclusion that the North/South divide in the NGO sector is as great a concern in the virtual world as it is in the physical one.[26] In 2015, *The Guardian* reported that "less than 2% of funds go directly to local NGOs, despite them taking the lion's share of the risk and often being better placed to deliver, according to aid insiders."[27] Consistent with scholarship on the subject, the authors writing for the *Caucasus Edition* exposed the extent to which the consolidation of funding at the hands of big consortia hollowed out the peacebuilding scene in the South Caucasus. If in the 1990s and 2000s multiple organizations and activist groups were collaborating across conflict divides, by the late 2010s the field had shrunk to a handful of NGOs that controlled nearly all the funding allocated to peacebuilding.

The criticism solicited mixed responses. Some donors and leaders of consortia received it as a politically motivated attack and threatened retaliation. Others were receptive and engaged with criticism constructively. Less predictably, when these few peacebuilders broke rank and walked away, they opened the floodgates. Many more authors and peacebuilders stepped out of the shadows of conformity and engaged in direct criticism of structures and organizations that had dominated the field with scant results to show for it. If in 2018 the critique of existing structures was exceptional, with all other articles of the *Caucasus Edition* staying within the confines of the liberal peace paradigm, the two 2019 issues of the journal were a significant departure, devoted to the exploration of postliberal approaches to peace[28] and to non-institutional peace activism in the South Caucasus.[29] With an increasing number of peacebuilders choosing to walk away from the existing structures to conceive and advance new and inclusive ones, the strategy started looking less like resignation or self-marginalization and more like an insurgency that aspires to transform the field.

By 2020 a series of program evaluations commissioned by the EU it-self corroborated many of the claims advanced by internal and external critics of the large consortia. As of this writing, the EU had restructured its funding mechanism for supporting the Nagorno-Karabakh peace process, breaking it down into a number of smaller and topical funding pools that prioritized the involvement of local NGOs. Moreover, as expectations concerning the oversaturation of the peacebuilding field due to large EU investment never materialized, a number of smaller donors returned, diversifying the funding pools.[30]

The Ones Who Walk Away to Stay and Fight

Conceptually, peacebuilding locates violence in systems and structures. Yet this recognition is rarely reflected in practices that grew preoccupied with quick-fix solutions that leave larger systems of violence intact. Richard Rubenstein urges peacebuilders to ensure that their initiatives, while addressing local problems, also contribute to system transformation and not maintenance.[31] This is easier said than done, however, as the systems are designed to keep challengers in check. The funding schemes and consortia privilege practitioners loyal to the existing structures and marginalize and exclude the rest.

In the South Caucasus in the 2000s and 2010s, most local NGOs and activists who chose to stay and fight, to work through the known funding schemes, had few avenues available through which they could enact meaningful change. Contending for anything more than a small subgrant required a demonstration of considerable bureaucratic capacity and a well-documented history of conformity to or acceptability by international and local authorities, reducing the field to a handful of international and local NGOs committed to system maintenance.

The smaller NGOs and activists who tried to stay and fight from within were relegated to competing with other local actors for 1% to 5% of total funds allocated for peacebuilding. The lucky few awarded the coveted consortia subgrants had to discover that the pennies they received against suffocating restrictions and rigid reporting requirements were enough for a small training or two, but not for bringing change.

Jemisin and Le Guin leave us with a profound dilemma when reckoning with systems of violence. Jemisin urges us to stay within the system and fight. A great many do. What they receive as starting capital, however, is not collegiality and support but "hunger games" that pit them against other local activists in an intense competition over an artificially limited funding pool. Walking away from the predictable but suffocating funding schemes and comfort of INGO patronage, leaving the marked path and pawing one's own way toward relative financial independence could be necessary steps to meaningful change.

Fundraising presents a strategic choice. One can map the available resources and tailor one's work to match these. Or one can start with a vision and create resources for advancing the dream. The former is the conventional yet the dead-end path. The latter is difficult but rewarding. Funding is never limited to the best-advertised schemes, which are structurally set to co-opt and control. Countless uninvolved yet persuadable foundations, individuals, and governments present a potential treasure trove of untapped but inconspicuous resources. The ones who walk away will face obstruction, exclusion, marginalization, and resource scarcity that will test their resolve. The battle will never stop. Those who succeed will find themselves showered with praise, attention, and funds, once more structured to co-opt. Some will divert, others will march on.

Le Guin's gambit is antithetical yet ethical. The ones who walk away abandon neither the child nor Omelas. They walk away to forge a path, for both themselves and their fight.

Notes

1. Ursula K. Le Guin, *The Ones Who Walk Away from Omelas* (Mankato, Minn: Creative Education, 1993).
2. N. K. Jemisin, *How Long 'til Black Future Month?* (New York: Orbit, 2018), 9.
3. Philip Gamaghelyan, *Conflict Resolution beyond the International Relations Paradigm: Evolving Designs as Transformative Practices in Nagorno-Karabakh and Syria*, Soviet and Post-Soviet Politics and Societies Series (Stuttgart: ibidem Press/Columbia University Press, 2017).

4. Anthony Giddens, *New Rules of Sociological Method: A Positive Critique of Interpretative Sociologies* (Stanford, CA: Stanford University Press, 1993).
5. Ivan Illich, ed., *Disabling Professions*, Ideas in Progress series (New York: M. Boyars, 1987).
6. Johan Galtung, "Violence, Peace, and Peace Research," *Journal of Peace Research* 6, no. 3 (January 1969): 167–191.
7. John J. Mearsheimer, *The Tragedy of Great Power Politics* (New York: Norton, 2003).
8. Laura Nader, *Harmony Ideology: Justice and Control in a Zapotec Mountain Village* (Redwood City, CA: Stanford University Press, 1991).
9. Jean-Paul Sartre, "Preface to Frantz Fanon's Wretched of the Earth," in The Wretched of the Earth by Frantz Fanon, translated by Richard Philcox (New York: Grove Press, 2004), xliii–lxii.
10. Julie Mertus and Jeffrey W. Helsing, eds., *Human Rights and Conflict: Exploring the Links between Rights, Law, and Peacebuilding* (Washington, D.C: U.S. Institute of Peace Press, 2006); Eileen Babbitt and Ellen L. Lutz, eds., *Human Rights and Conflict Resolution in Context: Colombia, Sierra Leone, and Northern Ireland* (Syracuse, NY: Syracuse University Press, 2009).
11. This fieldwork was undertaken in the summer of 2019.
12. Ned Lazarus, "Evaluating Peace Education in the Oslo-Intifada Generation: A Long-Term Impact Study of Seeds of Peace 1993–2010" (Ph.D. dissertation, American University, 2011).
13. Francis Fukuyama, *The End of History and the Last Man* (New York: Free Press, 2006).
14. Moisés Naím, "Democracy's Dangerous Impostors," *Washington Post*, April 21, 2007, https://www.washingtonpost.com/wp-dyn/content/article/2007/04/20/AR2007042001594.html.
15. Disclaimer: As of this writing, Philip Gamaghelyan is the lead methodologist for the EU-funded project led by the Imagine Center for Conflict Transformation, part of a consortium called EU4Peace and formed in 2020.
16. Between 2010 and 2018, 200,000 euros were distributed by the EPNK through small grants constituting less than 1.5% of the total sum of 14 million for the consortium. ATNP fared marginally better, distributing 5%, or 180,000 euros, between 2017 and 2019, compared with 3.8 million received by the consortium. For more details, see the report by NGO Monitor, "Facilitating Dialogue: EU-Funded NGOs and the Nagorno-Karabakh Conflict," October 2018, https://www.ngo-monitor.org/reports/facilitating-dialogue-eu-funded-ngos-and-the-nagorno-karabakh-conflict/.
17. Cavid Aga and Onnik James Krikorian, "No Hope for NK Peace? When Activists and Journalists Become Combatants," Meydan.TV, May 16, 2016, https://d9mc3ts4czbpr.cloudfront.net/az/article/no-hope-for-nk-peace-when-activists-and-journalists-become-combatants/.
18. NGO Monitor, "Facilitating Dialogue."
19. Disclaimer: Philip Gamaghelyan is a member of the editorial team at the *Caucasus Edition Journal* and serves as a board member at the Imagine Center for Conflict

Transformation, a regional peacebuilding organization that unites peacebuilders from many corners of the South Caucasus.

20. Oliver P. Richmond, "Resistance and the Post-Liberal Peace," *Millennium: Journal of International Studies* 38, no. 3 (May 2010): 665–692.

21. Vadim Romashov, Nuriyya Guliyeva, Lana Kokaia, and Tatia Kalatozishvili, "A Communitarian Peace Agenda for the South Caucasus: Supporting Everyday Peace Practices," in Challenging Gender Norms, Dealing with the Past, and Protecting the Environment: Community-Driven Conflict Transformation in the South Caucasus, ed. Philip Gamaghelyan, Sevil Huseynova, Maria Karapetyan, and Pınar Sayan. *Journal of Conflict Transformation* 3, no. 1 (2018): 157–173.

22. Milena Abrahamyan, Pervana Mammadova, and Sophio Tskhvariashvili, "Women Challenging Gender Norms and Patriarchal Values in Peacebuilding and Conflict Transformation across the South Caucasus," in Challenging Gender Norms, Dealing with the Past, and Protecting the Environment: Community-Driven Conflict Transformation in the South Caucasus, ed. Philip Gamaghelyan, Sevil Huseynova, Maria Karapetyan, Pınar Sayan. Special issue of *Journal of Conflict Transformation* 3, no. 1 (2018): 46–71.

23. Sona Dilanyan, Aia Beraia, and Hilal Yavuz, "Beyond NGOs: Decolonizing Peacebuilding and Human Rights," in Challenging Gender Norms, Dealing with the Past, and Protecting the Environment: Community-Driven Conflict Transformation in the South Caucasus, ed. Philip Gamaghelyan, Sevil Huseynova, Maria Karapetyan, Pınar Sayan. Special issue of *Journal of Conflict Transformation* 3, no. 1 (2018): 157–173.

24. Jeffrey Pfeffer and Gerald R. Salancik, *The External Control of Organizations: A Resource Dependence Perspective*, Stanford Business Classics (1978; Stanford, CA: Stanford Business Books, 2003).

25. Seth B. Cohen, "The Challenging Dynamics of Global North-South Peacebuilding Partnerships: Practitioner Stories from the Field," *Journal of Peacebuilding & Development* 9, no. 3 (2014): 69, https://doi.org/10.1080/15423166.2014.984571.

26. Michelle Shumate and Lori Dewitt, "The North/South Divide in NGO Hyperlink Networks," *Journal of Computer-Mediated Communication* 13, no. 2 (2008): 405–428, https://doi.org/10.1111/j.1083-6101.2008.00402.x.

27. Bibi van der Zee, "Less than 2% of Humanitarian Funds 'Go Directly to Local NGOs,'" *The Guardian*, October 16, 2016, https://www.theguardian.com/global-developm ent-professionals-network/2015/oct/16/less-than-2-of-humanitarian-funds-go-directly-to-local-ngos.

28. Philip Gamaghelyan, Sevil Huseynova, Christina Soloyan, Pınar Sayan, Sophio Tskvariashvili, eds. "Exploring Post-Liberal Approaches to Peace in the South Caucasus." *Caucasus Edition* 4, no. 1 (November 2019): 1–186.

29. Philip Gamaghelyan, Sevil Huseynova, Christina Soloyan, Pınar Sayan, Sophio Tskvariashvili, eds. "Activism and Peace in the South Caucasus" *Caucasus Edition* 4, no. 2 (2019): 1–155.

30. As of this writing in September 2020, the Imagine Center, where the author serves as a board member, has been invited to join a restructured EU funding scheme for

peacebuilding within the Nagorno-Karabakh conflict context. The new funding scheme now supports four consortia instead of one and has expanded the number of direct beneficiary organizations from six to dozens.

31. Richard E. Rubenstein, *Resolving Structural Conflicts: How Violent Systems Can Be Transformed*, Routledge Studies in Peace and Conflict Resolution (New York: Taylor & Francis, 2017), 136–139.

9

From Righteous to Responsive

Rethinking the Role of Moral Values in Peacebuilding

Reina C. Neufeldt

After about seven and half years doing peacebuilding fieldwork in a develop-ment and humanitarian relief organization, I returned to academia.[1] I found myself ruminating on those moments when I had a pit in my stomach about a project, but I had proceeded anyway. "Did we do the right thing?" I won-dered. Or sometimes the disturbing alternative: "Why did I think that was a good thing to do?" As I thought more carefully about the ethics of my choices in field practice, I heard people judging things as good and right in their work without recognizing they were making moral value judgments and likewise missing ethical discernment. In 2009, Tim Murithi voiced a similar assess-ment. He argued that the absence of an assessment of the ethical dimensions of peacebuilding contributes to its limited success.[2] This was a welcome shift that moved conversations in peacebuilding from focusing on tech-nical expertise to thinking about the values and ethical choices undergirding actions.[3] This volume's call to think further about ethics in action, to think about moral dilemmas and wicked problems, continues in this important vein. It deepens the conversation, highlighting the ways ethical dilemmas arise and are shaped by history, colonialism, and racism, as evident in the chapters here by Tony Gaskew and Liz Theoharis and Noam Sandweiss-Back.

There is a systematic gap in the field with respect to identifying, weighing, and discussing values and their effects on decisions and actions. Not that people are generally amoral or immoral, but at times our judgments about what was or is good, and the values upon which these judgments are made, themselves contribute to failures in peace work, for example, when the good intentions to help a war-displaced community access farmland leads to forced disappearances or fuels retaliatory violence, or when an emphasis on negotiating a peace deal means overlooking predatory sexual violence. These choices, made when confronting a dilemma, were informed by moral

Reina C. Neufeldt, *From Righteous to Responsive* In: *Wicked Problems*. Edited by: Austin Choi-Fitzpatrick, Douglas Irvin-Erickson, and Ernesto Verdeja, Oxford University Press. © Oxford University Press 2022. DOI: 10.1093/oso/9780197632819.003.0010 values, and they contributed to failure in peace work. This happens for people working on peacebuilding efforts in partnership with others outside of their community, as well as those working within their own community, although my work here focuses on lessons for people working with organizations on peacebuilding outside of their home communities. We need to not only recognize and think about ethical dilemmas and wicked problems but to think deeply about how we identify moral values that guide our actions and, most important, how we weigh and choose ethical actions.

Failure?

Before moving too far ahead, to what does "failure" refer? I could argue that a failure to improve things is a failure for peacebuilding and conflict resolution and transformation efforts. This is a point that Mary Anderson and Lara Olson have noted strongly as part of a multiorganizational effort to examine the effects of peace practice. They noted that, "so long as people continue to suffer the consequences of unresolved conflicts, there is urgency for everyone to do better."[4] This came from a collaborative research project into peace practice that built on the foundation of Anderson's "Do no harm" work noted in the introduction and continues its emphasis on impacts or outcomes. While I agree with this assessment, I am particularly concerned about failures that occur when interventions make things worse.

As individuals and organizations working in and on conflict, interveners can make things worse, which results in harm to people. Interveners can be responsible for the loss of life and can escalate conflict; outsiders can undermine local solidarities; insiders and outsiders can promise things that don't happen and increase people's cynicism and enmity. Interveners can divert resources and contribute to structural injustice, including those hoping to intervene and stop genocide, just as Mahmood Mamdani's analysis of the Save Darfur campaign illuminated an effort that was more about preserving a particular worldview than lives.[5] We can impose our values. We can use people for our ends of advancement, of realizing our own beliefs, or of what we think is a good outcome in a conflict. Each of these has contributed to failure at different times and places.[6]

On the converse side, what is success for a peacebuilder or conflict intervener? That is the challenge to which we must turn. My working definition of success is that ethical decisions emerge from a process of careful and open thinking that responds to the context and to the power inequities therein and contributes to collective human flourishing. In this light ethical decision-makers and actors embody excellence of character within relationships of care, accountability, and responsiveness. This is a working definition because it comes from my explorations of a limited set of moral theories: consequentialism, duty-based ethics, virtue ethics, ethics of care, and Ubuntu ethics. Note that these moral theories emerged in particular times and places and as such represent only some of the ways of understanding what makes something good or right or of how we can act ethically.

How Moral Values Contribute to Failure in Applied Peace and Conflict Work

These questions assume people want to do good work in the right ways and generally be ethical. I am not examining situations in which people deliberately aim to be unethical. While this problem also needs addressing, people generally enter the field of peace and conflict studies and social change efforts more broadly because they want to do good. The greater challenge for this group is to see the ways in which moral values and what we think is good can actually contribute to failure. There are at least three ways this failure can occur.

Moral and Ethical by Definition—That Is Good Enough

Thinking we are moral and ethical because of the work we do is an important contributor to failure. The editors of this volume allude to this same point in their introduction. In this line of thinking, our assessment of morality stops with our intentions. We made a value-based choice to work on conflict, to pursue peace, and now here we are engaging in action—no need to question further. This approach often infuses students' perspectives on the field of peace and conflict studies, as Agnieszka Paczyńska and Susan Hirsch explore later in this volume. Insidious effects, however, emerge from what Séverine Autesserre calls the "here to help" narrative.[7]

Autesserre argues that the ways in which international actors live and act in everyday work environments produce significant problems that make them counterproductive, ineffective, and inefficient.[8] One part of her analysis draws attention to the undermining role that a sense of moral superiority plays. Foreigners enter into a conflict for "moral reasons," "to help the host country and its citizens," and in so doing claim a moral high ground captured in the common saying "The hand that gives is always higher than the hand that receives."[9] There are two important dimensions of the "here to help" narrative. First, it separates interveners and expatriates into a "club" that is different from the local community. Second, expatriates then come to suggest variously that local communities lack capacity or are backward or incompetent or corrupt or only self-interested and doing this work for professional advancement or pay and not altruistic like the foreigners. Power inequalities further reinforce these divisions.

It is a problem when moral and ethical deliberations begin and end with the decision to intervene because everyday practices and attitudes evade scrutiny. Ask about international peacebuilders and you will hear some disturbing stories connected to the ways in which the international "club" acts.[10] Autesserre points out that it is not at all surprising that in contexts like this, local people frequently contest, resist, or reject international initiatives supposedly designed to help.[11] The focus here is on international peacebuilders, but the insights can also apply to mediators, advocates, and other peace workers within their own settings.

The Problem of Thinking We Know What Is Right

A second way moral values contribute to failure is when we think we know what is right and act upon our assumptions without deliberation. This can happen individually or as groups. The personal variant of this failure is commonly called dogmatism. Philosopher Anthony Weston speaks of dogmatism as one of three common substitutes or counterfeits for ethical thinking (the other two being relativism and rationalization).[12] Dogmatists know the answer to a moral question before it is raised; they cut off open and careful consideration of moral issues because they know what is right, regardless of the specific case or circumstances. Arguments then are simply attacks on another person or position regardless of what else might be morally salient. When we agree with the values that dogmatists hold we want to broadcast them (maybe retweet them), and when we disagree we think they should be silenced (close their Twitter accounts). In both situations, merely clinging to values without careful and open-ended thinking involves giving answers before we grasp the questions. For example, in chapter 3, Kirssa Cline Ryckman wrestles with questions of dogmatism and nonviolence, and in chapter 4, Ashley Bohrer compares two compelling stances that can be framed in dogmatic terms.

The second version of this problem manifests in faulty group decision-making processes. Irving Janis named this problem "groupthink."[13] Groupthink occurs when group pressures lead to a deterioration of "mental efficiency, reality testing, and moral judgment."[14] There are a variety of conditions that make this dynamic more likely to occur. What happens is that team members value unanimous agreement and group cohesiveness over open and reasoned debate or problem-solving. The group ignores contradictory information, becomes overconfident, and believes itself inherently moral. Outside opinions and groups are stereotyped. The decisions that result are irrational and problematic. Groupthink may be a problem particularly when people are located in a head office, making decisions under pressure and insulated from field complexities. Here again, the conviction that we know what is right is blinding.

The Problem of a Single Moral Value

A third way moral values contribute to failure is when we are guided by only one moral value in settings of conflict. We are trying to do good but see "good" only one way, without being consciously aware of it—a *my way or the highway* orientation. A story may help to help illustrate this point.

I worked as a peacebuilding technical advisor for a relief and development organization and focused on supporting high-quality peacebuilding programs, either stand-alone or integrated into development and emergency response initiatives. On one memorable occasion, I rushed off to provide technical input to a delegation of Burundians in the United States on a three-week training and planning visit.[15] The unique collaborative effort I was joining aimed to support the Burundian Catholic Church in its visioning and capacity for the pursuit of peace. There was a long history of support already in this relationship, as the Burundian Catholic Church responded to the violence of 1993 and thereafter.[16]

The planners of the workshop had spent most of their time working on spiritual reflection and then the components that involved exploring trauma, conflict transformation, and peacebuilding. They had not spent a lot of time thinking about how to develop the action plan that would meet funder and peacebuilding field standards. My aim—to infuse technical quality—was one I thought of as value-free or even as a positive contribution that the organizers and Burundian contingent should want. However, it produced conflict; organizers and participants were nonplussed as project planning techniques were suddenly injected into their carefully crafted space of spiritual and communal formation, growth, and transformation.

In the end, a project idea was developed and funded with significant reach in Burundi. The negative effect, however, was our valuing of efficiency and ends (understood as total number reached), which undermined Burundian and Catholic Church values of being in solidarity with those who are more disadvantaged, and subsidiarity, which refers to those closest to a problem taking the lead in resolving it. We undermined the ability of those in the United States to walk alongside the Burundian Catholic Church, and reinforced the value of technical expertise over deeply rooted spiritual and communal transformation.

In settings of conflict, particularly deep-rooted conflicts, values are part of the conflict. People fight to defend themselves against injustice, against oppression, or to fight for justice and freedom. Operating based on a narrow moral value—which we may not even recognize—contributes to our failure because we are unable to listen and to hear what values are important to multiple sides of the groups in conflicts. It can also contribute to distrust and schisms.

All three of these problems highlight the ways moral values can contribute to ethical failure. In each case, there is a general lack of thinking about and space for genuine deliberation on moral choices. If we hope to change the way we act, then we need to equip ourselves not only with lists of ethical principles but with new ways of thinking and acting together. The following are three initial responses to help us get started.

Response 1: Recognize Moral Values

The first response is to do better at recognizing moral values—that is, listening for those foundational ideas (things we consider important, of worth) upon which judgments are made. This is a skill worth practicing using sample text. I offer a short example from a UN Burundi Configuration document to illustrate.

The UN Burundi Configuration is a subgroup of the UN Peacebuilding Commission comprised of ambassadors representing various nations, such as Australia, Bangladesh, and Switzerland, as well as UN officials and representatives of regional and international bodies (e.g., African Union, World Bank). In 2015 the group penned an aspirational statement intended to send operational signals to Burundian leaders regarding what were then upcoming elections. At the time, the president of Burundi decided to stand for elections a third time; this was shortly after a failed coup attempt, and violence was escalating. The statement reads in part:

> The PBC Burundi Configuration highlights the importance of dialogue and reconciliation among all Burundians to address the root causes of the current crisis. It stresses the need to find a lasting political solution that ensures Burundi's hard gained progress in peace consolidation and peacebuilding.
> The PBC Burundi Configuration calls on all Burundians to urgently establish, through open dialogue and a spirit of compromise, the necessary conditions for the holding of free, transparent, credible, inclusive and peaceful elections.[17]

This statement contains a number of important claims about what is understood to be commonly agreed upon as good and right by the Burundian Configuration members—a particular set of actors, speaking into Burundian politics. In that first paragraph, we can see that nonviolent means of dispute settlement are valued as right and good in the phrase "dialogue and reconciliation" and, in the next paragraph, with the call for "open dialogue and a spirit of compromise." A negotiated political compromise is understood to be the right technical response, which actually relates to an underlying value, of political order and stability—values held dear by the Burundi Configuration and the UN more broadly. A necessary condition for political order is more carefully specified as "the holding of free, transparent, credible, inclusive and peaceful elections." In this phrase, we see several values espoused, including (again) nonviolence, participatory democracy, and transparency. These are understood to be good and right means. There is a statement about good ends in the document, not included in the excerpt, which refers to "sustainable peace and development." There is a strong statement about the wrong thing to do, which is to undermine or lose the "hard gained progress in peace consolidation and peacebuilding." Finally, there is an assertion that it is right and good for this set of international actors to speak into Burundian politics and to expect that Burundi will respond. In sum, nonviolent dispute settlement, political order and stability, participatory democracy, transparency, sustainable peace and development are some of the main moral values presented in the letter (there may be others).

One way to identify moral values is to listen to the reasons or justifications for why actors think it important to act, as well as what is judged right action. We can also identify moral values in the way a problem is framed. For example, the way I introduced the context of Burundi emphasized problematic political dynamics. The same occurs in the letter itself. This framing of the problem prioritizes political order as the most important moral good. While it is an understandable claim, it is also a limited moral claim that does not speak to other important components of what it means for Burundians to flourish, including personal well-being, recovering from trauma, or relational healing and nurturing community—all of which were important for the Burundi contingent.

Response 2: Listen beyond Ourselves

As we listen for moral values we need to listen not only to the strongest voices—like those speaking from the UN or a presidential palace, which are easier to hear—but also those that are quiet, disadvantaged, marginal. In his influential book *Conquest of America: The Question of the Other*, Tzvetan Todorov investigates what made it possible for European explorers to engage in mass extermination and conquest.[18] Values, he finds, formed one of three axes of alterity or "otherness" that helped unravel this puzzle. Values justified conquest for Christopher Columbus and Hernán Cortés in Mesoamerica; values were also at the center of Juan Ginés de Sepúlveda's arguments for why it was a Spanish right and duty to impose a Christian-informed understanding of good on others in a hierarchically organized world of superiority and inferiority.[19] Even the counterarguments of the Dominican bishop of Chiapas Bartolomé de Las Casas against Sepúlveda in a public debate relied on the assumption that the highest moral values were known by the Spanish.

There is an identification of "*my* values" as "*the* values" that are then imposed. Todorov asks, "Is there not already a violence in the conviction that one possesses the truth oneself, whereas this is not the case for others, and that one must furthermore impose that truth on others?"[20] This alludes to one of our challenges with state-led and UN pronouncements of what is good in peacebuilding, noted earlier.[21] In the letter I quoted, external actors within the system of states developed through colonization (and reinforced with decolonization) are declaring what is good for Burundi; there are power dynamics and histories to these claims that are complex and beg scrutiny. This does not operate only at the level of large-scale institutions but also at the personal level. My suggested response here is to recognize that we do not yet know what all the values are that we need to engage with—we might call this "values humility." In this response, it also becomes impossible to make all ethical decisions about peace work—work that involves communities— on one's own. Ethical deliberations are part of how we constitute ourselves within the relationships and communities in which we live and work.

Response 3: Embrace Plurality and Value Conflicts

When I speak of moral values, I am advocating an approach in line with Isaiah Berlin's value pluralism. Rather than argue for one moral theory (as many do in philosophy), Berlin argued that there are many genuine values to consider in ethical deliberation. This approach means that value clashes are inevitable. In Berlin's words, "We are faced with choices between ends equally ultimate, and claims equally absolute."[22] Mercy can clash with justice, the quality of life can clash with order, and so forth. Here we do not have a hierarchy of what is the better good.

To embrace value conflicts means bringing the both-and thinking of conflict resolution into creative collaborative decision-making in discerning ethical action.[23] We need to recognize that if we have a diverse set of values, or even a list of principles to guide our actions, values will come into conflict. Values, and principles derived from values, operate at a general level that require discernment for application; when they conflict, it is important to create space to think creatively and nondichotomously.

One example of creative thinking comes from a recent situation in a youth-community peacebuilding project with which I worked. This project, located in an urban area, involved NGO workers being pressured by an informal local leader to not proceed with the work unless they agreed to give some of the work to the leader's organization. Staff were physically intimidated and subject to regular threats. Here a youth peacebuilding effort seemed to be exacerbating conflict in the neighborhood. Valuing its open and transparent bidding process, the NGO took the informal leader's complaint seriously.

They did not want to be intimidated nor pay money as a bribe. However, they also valued the lives and security of their staff and the community. What could have been seen as a narrow dilemma around corruption or security was reframed. Staff members asked: What all is going on here? How else can we respond? By exploring the situation fully, it became clear that there was another value at play, which was more important to the informal leader: the value of respect. It was also clear that there had been some problems in the contracting process. The response then involved combining ways of demonstrating respect to the informal leader without the organization capitulating to specific demands and reshaping its working protocol in that community.

Reformulating how we operate, how we identify and analyze moral values, and the ways values focus our attention and actions and contribute to failure is important if peace and conflict resolution work is to be transformative. Whether working at home or abroad, as internationals or locals, there is a significant challenge before us. We need to recognize moral values, to listen and engage beyond ourselves, and embrace value conflicts that may even challenge our core values and deliberate on decisions within the relational context of our lives and work. It requires no less than working in a different way.

Notes

1. This chapter draws on material that was originally published in Reina C. Neufeldt, "When Good Intentions Are Not Enough: Confronting Ethical Challenges in Peacebuilding and Reconciliation," *Conrad Grebel Review* 36, no. 2 (2018): 114–132, and appears here with permission.
2. Tim Murithi, *The Ethics of Peacebuilding* (Edinburgh: Edinburgh University Press, 2009), 11.
3. Earlier works in conflict resolution that look at ethics include Herbert C. Kelman and Donald P. Warwick, "The Ethics of Social Intervention: Goals, Means, and Consequences," in *The Ethics of Social Intervention*, ed. Gordon Bermant, Herbert C. Kelman, and Donald P. Warwick (Washington, D.C.: Hemisphere, 1978), 3–36; James Laue and Gerald Cormick, "The Ethics of Intervention in Community Disputes," in *The Ethics of Social Intervention*, ed. Gordon Bermant, Herbert C. Kelman, and Donald P. Warwick (Washington, D.C.: Hemisphere, 1978); Wallace Warfield, "Is It the Right Thing to Do? A Practical Framework for Ethical Decisions," in *A Handbook of International Peacebuilding: Into the Eye of the Storm*, ed. John Paul Lederach and Janice Moomaw Jenner (San Francisco, CA: Jossey-Bass, 2002), 213–224.
4. Mary B. Anderson and Lara Olson, *Confronting War: Critical Lessons for Peace Practitioners* (Cambridge, MA: Collaborative for Development Action, Inc. Reflecting on Peace Practice Project, 2003), 10.

5. Mahmood Mamdani, *Saviors and Survivors: Darfur, Politics, and the War on Terror* (New York: Doubleday, 2010).

6. From chapter 7, reprinted with permission, Reina C. Neufeldt, *Ethics for Peacebuilders: A Practical Guide* (Lanham, MD: Rowman & Littlefield, 2016).

7. Séverine Autesserre, *Peaceland: Conflict Resolution and the Everyday Politics of International Intervention* (New York: Cambridge University Press, 2014).

8. Autesserre, *Peaceland*, 13.

9. Autesserre, *Peaceland*, 195.

10. See also Mary B. Anderson, "Can My Good Intentions Make Things Worse? Lessons for Peacebuilding from the Field of International Humanitarian Aid," in *A Handbook of International Peacebuilding: Into the Eye of the Storm*, ed. John Paul Lederach and Janice Moomaw Jenner (San Francisco, CA: Jossey-Bass, 2002), 225–234.

11. Autesserre, *Peaceland*, 13.

12. Anthony Weston, *A 21st Century Ethical Toolbox*, 3rd ed. (New York: Oxford University Press, 2013); Anthony Weston, *A Practical Companion to Ethics*, 4th ed. (New York: Oxford University Press, 2011).

13. Irving L. Janis, *Groupthink: Psychological Studies of Policy Decisions and Fiascoes* (1972; Boston: Houghton Mifflin, 1982).

14. Janis, *Groupthink*, 9.

15. A more detailed version of this story appears in Neufeldt, *Ethics for Peacebuilders*, 18–20.

16. In 1993, roughly 300,000 civilians were killed, half a million displaced, and a similar number became refugees in neighboring countries.

17. From the United Nations Secretary General, "Statement by the Burundi Configuration of the UN Peacebuilding Commission," May 15, 2015, https://www.un.org/sg/en/content/sg/note-correspondents/2015-05-15/statement-burundi-configuration-un-peacebuilding.

18. Tzvetan Todorov, *Conquest of America: The Question of the Other*, trans. Richard Howard (Norman: University of Oklahoma Press, 1999).

19. Todorov, *Conquest of America*, 151–153.

20. Todorov, *Conquest of America*, 168.

21. While I use Todorov's work for peacebuilding, others who have explored some of this terrain in contemporary peacebuilding include Roland Paris, "International Peacebuilding and the 'Mission Civilisatrice,'" *Review of International Studies* 28, no. 4 (2002): 637–656; Meera Sabaratnam, "History Repeating? Colonial, Socialist and Liberal Statebuilding in Mozambique," in *Routledge Handbook of International Statebuilding*, ed. David Chandler and Timothy D. Sisk (London: Routledge, 2013), 106–177.

22. Isaiah Berlin, *Liberty* (New York: Oxford University Press, 2002), 213–214.

23. Anthony Weston, *Creative Problem-Solving in Ethics* (New York: Oxford University Press, 2007).

SECTION III
SYSTEMS AND INSTITUTIONS

Here we take a fresh look at some of the ethical dilemmas that surround nonviolent civilian protection, human rights advocacy, the protection of the rights of perpetrators of violence and parties in armed conflict, the prevention of genocide and mass atrocities, transitional justice, and international sanctions, asking *How do powerful institutions shape the terrain of peacebuilding, both as catalysts for change and as generators of various obstacles and challenges?*

We have not divided chapters based on their domestic or international focus. It is important to emphasize that so-called domestic issues of peace and conflict in societies are inextricably bound to issues of peace and conflict everywhere in the world. We have avoided a domestic/international dichotomy precisely because we consider the United States to be just another case study in our global survey of wicked problems and ethical dilemmas.

We thought it was fitting, therefore, to conclude with observations on the dilemmas arising in two powerful northern institutions that have significant influence in how peacebuilding and conflict resolution are practiced worldwide: nongovernmental organizations, and the academy. The final two chapters make a forceful call for these two institutions to develop a robust ethical code of conduct, to uphold the rights and moral standing of research subjects (including vulnerable civilians), and to address asymmetric power relations inherent in both institutions' work. Indeed, how we actually educate future peacebuilders is itself a key ethical question, one that may have long-term effects on their understanding of and conduct in the world. Even if peacebuilders, activists, academics, and students can identify criteria for deciding when to respond to violence against civilians, there is still the question of how to respond. We hope that upon finishing this book, the reader might reflect on the ethical dilemmas surrounding knowledge about violence and peace in conflict zones, no matter where we are.

10

Dilemmas in Action Where Rule of Law Conflicts with Justice

Deena R. Hurwitz

On one of my last days in Palestine (West Bank), a close Palestinian friend and colleague asked if I would help him with something. I was there for an academic conference and to meet town leaders of Battir, with whom I was working pro bono on their (successful) petition to be designated a UNESCO World Heritage Site. My friend asked if I would take clothing and books back to the United States for a student who had stayed with his family while interning at the conflict resolution center that he directs. Returning from a short trip to Jordan, the student was denied reentry by Israeli military authorities and forced to return to the United States without her belongings.

Every person, almost without exception, is questioned at departure by Israeli airport security (usually young army soldiers or reservists). Standard questions: Where have you been? What were you doing? Did you pack your bags yourself? Did anyone give you something to take? Palestinians and other Arabs, including Arab Americans, and anyone identified as having been in the Occupied Palestinian Territories, are routinely subject to more pointed questioning and their bags are searched.[1] I have worked in Israel and Palestine for four decades and am intimately familiar with the varied treatment people receive on the basis of national origin and guilt by association. Sometimes my Jewish name elicits a perfunctory recitation of the questions; sometimes it is more aggressive, for example when it becomes clear that I have been visiting people's homes or working in Palestine. In all cases, an affirmative reply to the last question—Did anyone give you something to take?—guarantees prolonged questioning, having your bags vigorously searched and possibly confiscated. You may miss your flight. More significant, you may be threatened until you identify the person who gave you the items, setting up a high risk that *they* will be interrogated or arrested or perhaps have their property confiscated or destroyed.

Deena R. Hurwitz, *Dilemmas in Action Where Rule of Law Conflicts with Justice* In: *Wicked Problems*. Edited by: Austin Choi-Fitzpatrick, Douglas Irvin-Erickson, and Ernesto Verdeja, Oxford University Press. © Oxford University Press 2022. DOI: 10.1093/oso/9780197632819.003.0011

It was understood that I could not say where I got the clothes and books. Was I being asked to lie, or to break a law? What law was that? How does law function between a sovereign state that holds the power and those people living under their occupation, who have no power? What purpose does it serve for them?

Such questioning may seem reasonable to many, living as Israelis and others do under a constant fear of terrorism. Certainly, Israel is not the only country that practices racial profiling, and the public order depends on honesty and transparency. But we should not unconsciously accept the underlying premises of discriminatory or inequitable actions or policies just because a government says it is in the interest of national security.

This situation is not uncommon. Human rights advocates in the field, even government officials at times, are asked by vulnerable local colleagues to surreptitiously carry out documents, letters, notes, photographs, even a bag of clothes. (This is *not* about weapons or other illicit items.) Human rights lawyers work in places where the rule of law is compromised; it may be a country ruled by a dictator or an ostensibly liberal democracy governed by national security (or military) law. How forthright should one be with a rights-abusing state when interrogated upon entrance, exit, or any number of other encounters? Should one answer truthfully as a matter of principle, although it may compromise the safety of colleagues, clients, friends? Should one respond with false information? Be evasive or give partial information? Or should one avoid being put in that position? Is that always possible, or practicable? What would that even mean?

In this chapter, I tease out some of the ethical dilemmas that arise when lawyers encounter the manipulation of law for political ends. They are dilemmas of personal conscience and of professional responsibility, of moral accountability and complicity—and how that is defined, by and for whom. Nearly all of the contributions in this book engage with these themes. Ernesto Verdeja, for example, discusses the centrality of power to ethical dilemmas; being both constitutive and regulative, power frames the process of ethical decision-making.

I believe that it is incumbent on lawyers (and others) to continually question the law. Legal professionals are tasked with upholding the rule of law, but what if the law is unfairly skewed to disadvantage certain groups of people, and upholding it means being complicit in injustice? Beyond principles, do lawyers have an obligation—legal or moral—to challenge laws, policies, rules, or orders that they believe are illegitimate or unjust? What should they do when faced with unethical situations? What choices and what duties do lawyers have to moral accountability, particularly when it calls for action beyond the borders of the legal system? Professional ethics calls on lawyers to exercise lawful means to repeal or amend laws that are unjust or do not respond to the needs of society.[2] Yet what if you can't improve or prevent injustice by lawful means? What if the state is the lawbreaker? Does this heighten the moral obligation for lawyers?

Martin Luther King Jr. wrote that an individual who breaks a law that their conscience tells them is unjust, and willingly accepts punishment, is expressing the highest regard for the law.[3] But we should ask: Under what structures or circumstances is punishment or reward given?

This is not unique to lawyers; journalists, anthropologists, health workers, peacebuilders, and conflict mediators, among others, face these dilemmas in various ways. However, as a rule of professional conduct, lawyers are not supposed to knowingly make false statements of fact. In the United States, misconduct under the *Model Rules of Professional Conduct* includes engaging in conduct involving dishonesty, fraud, deceit, or misrepresentation.[4] On the other hand, generally speaking, the rules of professional conduct are rules of reason—meaning they are situational. And they are almost exclusively geared to client representation and courts in an adversarial setting. Beyond deliberate false statements in that context, the scope of obligations to truth and integrity are less clear. As ethical rules of reason, there is a certain amount of "permissible dissembling" allowed to identify wrongdoing or avoid further immoral acts.[5] So when, if ever, is it acceptable to disobey or circumvent the law? When, if ever, is there a moral imperative to act outside the law?

Lawyers' Role and Responsibility with Legal Power

A distinctive feature of ethics in any profession is that it addresses unequal encounters of two presumably moral persons, groups, or even institutions.[6] That imbalance is compounded by the structures of political and socioeconomic power, the rule of law, and access to justice. Legal professionals serve essential roles in the mediation of the social order, the rule of law. In providing access to justice, they interpret the law. Whether the law is just, transparent, and equitable, or whether it is unjust, discriminatory, or repressive, those who interpret, administer, and enforce it bear the responsibility of that power. With that responsibility comes an obligation to ethical practice.

Ethical practice involves critical consciousness and vigilance in the routine exercise of the rule of law. Why vigilance? Because legal systems, like political systems, are change-averse. Conventional lawyering conceives of law as a neutral corpus, and the profession as an impartial gatekeeper of the rule of law. Social justice lawyering, though, asserts that the law is not and has never been neutral. The law reinforces political and socioeconomic interests, and it is wielded in the service of those interests irrespective of ethical considerations.

Lawyers engage in social justice lawyering within the framework of the legal system in different ways. Some lawyers provide a skilled and zealous defense for vulnerable clients, while serving the public interest. Austin Sarat and Stuart Scheingold introduced the term "cause lawyering," referring to a form of moral activism using legal means to achieve greater social justice.[7] Cause lawyers seek to advance social, economic, and political change—a focus that transcends client representation.

Social justice lawyers connect law and reality with the goal of exposing and undoing structural, procedural, and psychological exploitation. Access to justice is historically engineered to favor particular groups with power. Thus, ethical practice necessitates questioning, and at times challenging, the rule of law to confront, alleviate, or eradicate injustice.

As an instrument of social control, law legitimizes the exercise of coercion and countenances systems of authority and privilege.[8] Recognizing that the legal system is implicated in the systemic exclusion of oppressed groups, and challenging the neutrality of the legal process, raise difficult questions about the limits of a lawyer's role.[9] Are we obligated to achieve social change within the framework of the law? Is it always possible to do that? If one accepts that the legal system is not a level playing field, that it is not a neutral set of rules, then to act as if it were is itself a moral choice regarding one's relationship to individuals, the community, and society as whole. There are ethical implications in functioning as if lawyers are objective and, by implication, apolitical professionals following the law. For that matter, is it always possible to be apolitical? Why is neutrality considered a positive attribute? Not that it is unnecessary, but to what degree and toward what end should one aspire to be neutral? Subjectivity is not a weakness; it can be a strength, depending on the basis for one's position.[10]

Whether to participate, withdraw, or disobey implies that the lawyer has a choice. In some cases, lawyers may not have the choice. There the dilemma may be whether to comply with an unethical regulation, policy, or order, or refuse and risk the consequences of what might be determined lawbreaking by the legal (and/or political) system. All actions involve costs for the individual. As Hannah Arendt wrote, "Those who choose the lesser evil forget very quickly that they chose evil." The question is, who bears what level of cost: the individual (lawyer), the at-risk people, victims of rights violations, the normative foundations of society? Or all of the above, and to what degree?

Ethical Rules and Considerations of Legal Practice

Among the essential duties of the legal profession are zealous advocacy, confidentiality, avoidance of conflicts of interest, improving the law, and upholding the reputation of the profession. The prevailing view is that zealous advocacy "stops at the 'bounds of the law.'"[11] Working for legal and structural change is advanced (some would argue *required*)—within the system, for example as a civil servant, legislator, judge, prosecutor, or impact litigation attorney. The law also can be challenged through investigation, exposition/reporting, and shaming. Moving toward the bounds, one might refuse to participate in or cooperate with the enforcement or implementation of unethical policies or laws, for example by boycotting a proceeding, whistleblowing, or resigning one's position.

Yet the ethics rules impose conflicting duties on lawyers with respect to working within the law. The codes state that lawyers have a "duty to uphold legal process" and to show respect for the law by obeying it.[12] On the other hand, the directive to lawful means is in tension with the assertions that "each lawyer must find within [their] own conscience the touchstone against which to test the extent to which [their] actions are ethical."[13] A lawyer's professional duties include "when necessary . . . challeng[ing] the rectitude of official action"[14] and must be informed by "personal conscience," not merely what is prescribed in the codes of ethics.[15] Indeed, the U.S. Supreme Court noted, "A bar composed of lawyers of good character is a worthy objective, but it is unnecessary to sacrifice vital freedoms in order to obtain that goal."[16] The freedom to criticize and resist the state's authority is not taken away from lawyers by their choice of profession.[17]

Sometimes efforts to improve the law or prevent injustice by lawful means are ineffective or unfeasible. Hofstra Law professor and director of the Institute for the Study of Legal Ethics Ellen Yaroshefsky asserted that "lawyers may or even must disobey unjust laws or take reasonable action to restore respect for law and fundamental human rights in an unjust system of laws. Within this view, nullification of unjust laws through individual action that upholds underlying legal values is acceptable."[18] Legal ethics scholar and Georgetown Law professor David Luban pointed out that "because lawyers are often better positioned than nonlawyers to realize the unfairness or unreasonableness of a law, lawyers often should be among the first to violate or nullify it."[19] And yet one could argue that lawyers are less well positioned because they often cannot see the unfairness, bound as they are to a legal system, structure, or set of interests that can just as easily obstruct reason or vision.

Ethical Action as a Principle of Human Rights Practice

The rules of professional conduct offer little to guide a lawyer intent on ethical practice when the state is the lawbreaker or lawmakers are manipulating the law to particular political ends, as with the Torture Memos, which justified the U.S. government's use of torture against detainees in the "war on terror" during the George W. Bush administration.[20]

To monitor and document human rights abuses, lawyers must get into and out of conflict zones, and they have to trust and be trusted by human rights defenders on the ground. What ethical dilemmas arise in this context? First among many, whether to be forthright with the rights-abusing state about the purpose for entering. The trade-offs are evident: one might not get in; one might get in and have one's movement controlled, be followed covertly or overtly, be arrested; or one's contacts might face retaliation, arrest, and worse.

As the editors of this volume assert in the introduction, the "Do no harm" principle is a starting point, not the highest standard. A significant theme throughout this book, "Do no harm" must begin with an understanding of the history and context of the conflict zones in which one is working. This means being conscious of one's privilege and power merely by virtue of being able to leave. Or choosing to stay in a hotel where the water is hot and electricity runs all day. How to go in and out of an area with respect for the conditions on the ground and the danger under which people live. Reina Neufeldt, Elizabeth Hume and Jessica Baumgardner-Zuzik, alicia sanchez gill and her co-narrators, and Minh Dang, among others, note that for a practitioner, a self-reflective understanding of one's position in relation to the local and global political economy and context is key. When we work in the field, wrote Sara Roy, senior research scholar at Harvard's Center for Middle Eastern Studies, our role is to "render visible the complexities of . . . life and, in so doing, provid[e] a more differentiated understanding of the forces that shape it. I do this while acknowledging that all interpretations can—and should—be challenged, recalling what Paul Ricoeur once wrote, 'Neither in literary criticism, nor in the social sciences, is there . . . a last word.'"[21]

* * *

Back to my situation with the Israeli airport security. When asked if I was given anything, I did not answer affirmatively with regard to the student's clothes and books. Was I ethically bound to tell the Israeli military that these clothes belonged to an American peace studies student whom they deported rather than allow her to return to her internship in the West Bank, and were entrusted to me by Palestinian friends who could not take them out themselves? Did I lie, or dissemble the truth? Do ethical principles demand an honest response to a policy that itself is unethical and violates human rights and would place an innocent person in danger? Protecting people at risk by being conscious of their conditions, carefully choosing what is said to state authorities, is a moral obligation of human rights work.

Should we decline to help an at-risk colleague because it might subject us to probing questioning by authorities? It might go something like this: "I'm sorry, I know you live in a war zone, under occupation. However, I can't risk having my bags rifled or being detained and possibly missing my flight by carrying something for you. Thank you, though. It was very important for our work that we met." There is an ethical choice in declining such a request from someone who would likely be racially profiled, searched, or prevented from exercising their right to freedom of movement.[22]

Is it unethical to respond to rules or state agents with partial truths in order to protect vulnerable persons? Do U.S. lawyers owe a duty of candor to a foreign government, especially one that violates human rights? Is this civil disobedience? Is a moral choice at play where the law or legal system is discriminatory, rigged against a particular group, or where unrestrained "national security" is intended to silence all questions or challenges? When governments lie, present incomplete truths or misinformation, or withhold information, why are lawyers required to accept them by following the law? To act with noncompliance or civil disobedience from the assumption— let alone evidence—that they are untruths is deemed unethical and unlawful. But to not act is to make a choice of complicity. There is no legal or ethical prerogative to be a bystander to injustice. There is an ethical imperative to be a witness to injustice, and being a witness presents individuals with a call to action.

When Humanitarian Assistance Is a Subversive Act or a Crime

I want to discuss another case that highlights certain dilemmas faced by prosecutors who apply laws that may allow for discretion, may be ambiguous, or may be clear but unjust.

Along the U.S.-Mexico border, volunteer groups for years have been leaving food, water, and medical supplies, hoping to mitigate for migrants the lethality of the scorching desert.[23] The humanitarian crisis became even more acute when, under the guise of national security, the Trump administration cut off legal avenues for refugees to enter the United States. Migrants attempting to cross between border checkpoints face getting lost in the desert, suffering from dehydration, and dying.[24] Between 2000 and 2018, the remains of over three thousand people have been found in the Sonoran Desert alone, although experts estimate that over ten thousand have died while attempting to cross.[25] Some human rights defenders who undertake this kind of work are trained to follow written protocols and procedures to make sure their aid is effective, responsible, and legal.[26]

In January 2018, Scott Warren of the humanitarian aid group No More Deaths/No Más Muertes was arrested and charged with harboring "aliens" and conspiracy. According to his indictment, giving migrants in need food, water, beds, and clean clothes are federal crimes for which he faced twenty years in prison. The first trial ended in June 2019 in a hung jury; eight of the twelve jurors voted for acquittal. Federal prosecutors dropped the conspiracy to transport or shield charge and proceeded with a retrial on two counts of harboring an undocumented immigrant, for which Warren still faced up to ten years in prison.[27] During the six-day trial in November 2019, prosecutors successfully asked the judge to bar the defense from mentioning President Trump or his policies. On November 20, after some two hours of deliberation, a jury found Warren not guilty. He declared, "The government has failed in its attempt to criminalize basic human kindness."[28]

What moral compass, let alone legal reasoning, points toward prosecuting this kind of act? Prosecutors have significant discretion as to where they devote time and resources.[29] The *Model Code* recognizes that the responsibility of a prosecutor "differs from that of the usual advocate; his duty is to seek justice, not merely to convict."[30] Imagine the impact of a prosecutor adopting even one of Verdeja's four principles underpinning prevention: dignity, empathy, universal responsibility, and "Do no harm."

Fundamentally, the protection of individual rights from government abuse is a key aspect of our legal and political systems. Professor Luban admonished, "Prosecutors shouldn't exploit the power imbalance. . . . [They are] the gatekeeper[s] of the system. . . . The prosecutor's conscience is the invisible guardian of our rights, just as the defense lawyer is the visible guardian."[31]

Conclusion

These problems may be ethical trade-offs, but politicization of moral choice is also at play. It is the appropriation of ethics in the name of—or as—the rule of law to ensure that certain acts of social justice are costly and politically untenable. In particular, when national security is used to justify policies or regulations that limit or violate human rights, conscientious citizens must choose between being a bystander to or challenger of the abuse of power. Lawyers must be conscientious citizens with an elevated sense of moral responsibility, as interpreters and enforcers of the rule of law.

The notion of neutrality assigns the law certain values that must be continuously interrogated. As Sara Roy asserts:

> Any claim to neutrality or, for that matter, objectivity is, in my experience, nothing more than calculated indifference. . . . My commitment is to accuracy—to representing the facts to the best of my ability—not neutrality or objectivity; neither is possible in any event. Neutrality is often a mask for siding with the status quo and objectivity—pure objectivity—does not exist and claiming it is dishonest.[32]

Interpreting the law as if it were neutral may admit one to an elite club. But the consequences of such a value system include contributing to a social order that views challenging injustice unethical or illegal. Where social justice is the objective, moral activism is a catalyst. Former Harvard Law dean and professor Martha Minow submitted, "Our legal system needs people who are willing to break the law for political reasons. Such people provide one of the checks on the system's otherwise effective and often well-placed curbs on social change."[33] Liz Theoharis and Noam Sandweiss-Back, among others in this book, understand that it will take a mass social awakening to shift out of, and to undo, these systems of violence. As Frederick Douglass famously said, "Power concedes nothing without a demand; it never has, and it never will."

If illegality and injustice are allowed to persist and become not only institutionalized but ingrained in our assumptions and our worldview, how does one repair the damage, and when? Is it even possible to do so?

There is a fundamental asymmetry that speaks to the law and how it is applied for different people, and there is a moral inequity between a humanitarian worker or human rights defender and someone like John Yoo, who wrote the Torture Memos. Despite being the intellectual author of such a perversion of the law, Yoo resumed his position as a professor at a prestigious law school and went on to advise the government on ways to circumvent the law to impose its preferred policies. Meanwhile, lawyers, activists, scholars, government employees, and others who act to protect and defend rights are punished for breaking the law. What does this reveal about our system of justice, how individual lawyers must behave, and the kinds of decisions they are called on to make?

Ethical dilemmas are common, normative, and inevitable. When we believe we are constrained by, bound to, a set of rules—right or wrong—we are in danger of losing our moral compass. We have a professional obligation to use our knowledge and expertise to think critically and continually interrogate the basis for and the purpose of the rule of law. Is the law doing what it is intended to do? Is it fully consistent with human rights and fundamental freedoms for all persons? Or is the tail wagging the dog? Everything we do involves moral choice, and we must consciously consider the impact of the choices we make. To join a cross-cutting refrain of this book: not acting is a choice to accede to and even participate in injustice.

Notes

1. It must be acknowledged that just getting to the airport often is an ordeal of time and humiliation for Palestinians, and they are subject to discrimination at all border crossings, not just the airport. Despite changes resulting from a 2007 petition filed by

the Association for Civil Rights in Israel (ACRI), the fundamental flaw persists: the treatment of national origin as a criterion for deciding the scope of a security check, and of all Arab citizens as potential threats simply because they are Arab. ACRI. "Situation Report." 2014.

See also, Peter Beinart, "I Was Detained At Ben Gurion Airport Because Of My Beliefs," The Forward, Aug 13 2018, available at: https://forward.com/opinion/408 066/peter-beinart-i-was-detained-at-ben-gurion-airport-because-of-my-beliefs/ (last visited Oct 27, 2021); Alexandra Goss, "Banned from the Only Democracy in the Middle East: Targeted Exclusion at Israel's External Border Crossings" (2016). Pomona Senior Theses. Paper 166, available at: http://scholarship.claremont.edu/po-mona_theses/166 (last visited Oct. 29 2021) (study drawing on cases of 110 U.S. citizens banned or denied from external Israeli border crossings between 1987-2015, and finding that denial is often associated with Palestinian Americans, Arab Americans, Muslim Americans, Black Americans and Americans who personally identify as activists or are considered by border officials to be so); Gabe Stutman, "Activist detained at Ben Gurion: Once I said the word 'Arab,' everything changed," The Times of Israel, Apr 19, 2019, available at: https://www.timesofisrael.com/activist-detained-at-ben-gurion-once-i-said-the-word-arab-everything-changed/ (last visited Oct. 29, 2021); George Khoury, 'This is our Israel, this is for the Jews. No Palestinian should come to Israel': A Palestinian-American's story of being detained at Ben Gurion airport," Mondoweiss, July 29 2015, available at: https://mondoweiss.net/2015/07/pale stinian-americans-detained/ (last visited Oct. 27, 2021).

2. American Bar Association, Model Code of Professional Responsibility, 1969, EC-8-2, https://www.americanbar.org/content/dam/aba/administrative/professional_res ponsibility/2011build/mod_code_prof_resp.pdf, hereafter Model Code 1969.

3. Martin Luther King Jr., "Acceptance Address for the Nobel Peace Prize," Oslo, December 10, 1964, Martin Luther King, Jr. Research and Education Institute, Stanford University, 82. https://www.nobelprize.org/prizes/peace/1964/king/facts/

4. American Bar Association, Model Rules of Professional Conduct, 2000, 8.4(d) and (g), hereafter Model Rules 2000. https://www.americanbar.org/groups/professional_re-sponsibility/publications/model_rules_of_professional_conduct/

5. American Bar Association, "When Is It OK for a Lawyer to Lie?," December 2018. https://www.americanbar.org/news/abanews/publications/youraba/2018/december-2018/when-is-it-okay-for-a-lawyer-to-lie--/.

6. Monroe H. Freedman and Abbe Smith, Understanding Lawyers' Ethics, 3rd ed. (Newark, NJ: LexisNexis, 2004), 7.

7. Austin Sarat and Stuart Scheingold, Cause Lawyering: Political Commitments and Professional Responsibilities (Oxford: Oxford University Press, 1998).

8. Robert M. Cover, "The Folktales of Justice: Tales of Jurisdiction," Capital University Law Review 14, no. 2 (1985): 180.

9. Kathryn Abrams, "Lawyers and Social Change Lawbreaking: Confronting a Plural Bar," University of. Pittsburgh Law Review 52, no. 4 (1991): 755.

10. See Sara Roy, Failing Peace, Gaza and the Palestinian-Israeli Conflict (London: Pluto Press, 2007), xvi.

11. Ellen Yaroshefsky, "Military Lawyering at the Edge of the Rule of Law at Guantanamo: Should Lawyers Be Permitted to Violate the Law?," *Hofstra Law Review* 36, no. 2 (2007): 581.

12. *Model Rules 2000*, Preamble para. 5; *Model Code 1969*, EC 1-5.

13. *Model Code 1969*, Preamble.

14. *Model Rules 2000*, Preamble, para. 5.

15. *Model Rules 2000*, Preamble, para. 7.

16. *Konigsberg v. State Bar of Cal.*, 353 U.S. 252 (1957), 273.

17. Robert M. Palumbos, "Within Each Lawyer's Conscience a Touchstone: Law, Morality and Attorney Civil Disobedience," *University of Pennsylvania Law Review* 153 (2005): 1087.

18. Yaroshefsky, "Military Lawyering at the Edge of the Rule of Law at Guantanamo," 581.

19. David Luban, "Legal Ideals and Moral Obligations: A Comment on Simon," *William & Mary Law Review* 38, no. 1 (1996):259.

20. See David Luban, "The Defense of Torture," review of *War by Other Means: An Insider's Account of the War on Terror*, by John Yoo, *New York Review of Books*, March 15, 2007; John Yoo, "Bybee Memorandum," U.S. Department of Justice, memorandum from Jay S. Bybee, assistant attorney general, to Alberto R. Gonzales, counsel to the president, Re: Standards of Conduct for Interrogation under 18 U.S.C. sec.2340–2340A, August 1, 2002, https://www.justice.gov/olc/file/886061/download; Andrew Cohen, "The Torture Memos, 10 Years Later," *The Atlantic*, February 6, 2012.

21. Sara Roy, *Hamas and Civil Society in Gaza, Engaging the Islamist Social Sector*, 2nd ed. (Princeton, NJ: Princeton University Press, 2014), 18.

22. Beyond declining to assist and not complying with authorities, there are other options. When it can be done without compromising those you are helping, you might choose to "speak truth to power"—answering candidly out of a conviction that refusing to be intimidated and assuming the consequences yourself can challenge the very basis (and objective) of the injustice, turning it back on the authorities. It might go something like this: "Yes, actually, these clothes belong to an American peace studies student you deported for no reason other than she was living and working with Palestinians." It is a choice of which battles to fight, and where.

23. Amy Goodman and Denis Moynahan, "Humanitarian Groups Continue Work at U.S-Mexico Border, Despite Risk of Prosecution," *Rabble.ca*, August 22, 2019, https://rabble.ca/columnists/2019/08/humanitarian-groups-continue-work-us-mexico-border-despite-risk-prosecution.

24. Wendy Feliz, "Providing Humanitarian Aid to People Crossing the Border Should Never Be a Crime," *Immigration Impact*, June 17, 2019.

25. Humane Borders/Fronteras Compasivas, "Migrant Death Mapping, Migrants Deaths, Rescue Beacons, Water Stations 2000–2018," accessed July 23, 2020, https://humaneborders.org/migrant-death-mapping/; Geoffrey Alan Boyce, Samuel N. Chambers, and Sarah Launius, "Bodily Inertia and the Weaponization of the Sonoran Desert in US Boundary Enforcement: A GIS Modeling of Migration Routes through Arizona's Altar Valley," *Journal on Migration and Human Security* 7, no. 1 (March 6, 2019), doi: https://doi.org/10.1177/2331502419825610.

26. Bob Ortega, "Trial Begins for No More Deaths Volunteer Who Aided Migrants," *CNN Investigates*, June 3, 2019.

27. Sanjana Karanth, "Federal Prosecutors to Retry Border Humanitarian Volunteer for Helping Migrants," *Huffington Post*, July 2, 2019.

28. Teo Armus, "After Helping Migrants in the Arizona Desert, an Activist Was Charged with a Felony. Now, He's Been Acquitted," *Washington Post*, November 21, 2019. In 2018, over 4,500 people were federally charged for bringing in and harboring migrants—a more than 30% increase since 2015. The greatest rise followed former attorney general Jeffrey Sessions's order to prioritize harboring cases. See Lorne Matalon, "Extending 'Zero Tolerance' to People Who Help Migrants along the Border," *All Things Considered*, National Public Radio, May 28, 2019, https://www.npr.org/2019/05/28/725716169/extending-zero-tolerance-to-people-who-help-migrants-along-the-border.

29. Freedman and Smith, *Understanding Lawyers' Ethics*, 306n6.

30. *Model Code 1969*, EC 7-13; see also David Luban, "The Conscience of a Prosecutor," *Valparaiso University Law Review* 45, no. 1 (2010): 1–31.

31. Luban, "The Conscience of a Prosecutor," 14.

32. Roy, *Failing Peace, Gaza and the Palestinian-Israeli Conflict*, xvi.

33. Martha Minow, "Breaking the Law: Lawyers and Clients in Struggles for Social Change," *University of Pittsburgh Law Review* 52, no. 4 (1991): 741-743.

11

Establishing an Ethics of
Post-Sanctions Peacebuilding

George A. Lopez and Beatrix Geaghan-Breiner

In this chapter we provide an initial schematic for establishing a post-sanctions ethics of peacebuilding. George comes to this question after thirty years of research and practice work in both sanctions and post-violence peacebuilding. He believes ethical dimensions of sanctions have been underexplored and he seeks correctives. Beatrix approaches this ethical dilemma as an undergraduate student studying U.S. foreign policy. Her research on the impacts of Venezuelan sanctions, undertaken as a Laidlaw Scholar at Columbia University, has introduced her to the humanitarian costs of economic statecraft. She joins George in the articulation of a policy framework that aims to redress sanctions-inflicted harm on innocent civilians.

Building from international law discussions about forceful humanitarian intervention as "the responsibility to protect," international relations experts focused on "the responsibility [or the imperative] to rebuild" war-torn nations. For their part, ethicists, with a conscious link to the more traditional just-war categories *jus ad bellum* and *jus in bello*, defined and then studied this undertaking as *jus post bellum*.[1] Numerous sanctions scholars and ethicists have condemned sanctions using *in bello* criteria and increasingly have labeled sanctions as economic war. But they develop neither principles nor practices that provide an ethical response to post-sanctions situations. If *post bellum* peacebuilding warrants a responsibility to rebuild, what responsibility does post–economic sanctions peacebuilding require? It is this ethical dilemma our chapter sets out to explore.

We examine this question by documenting how sanctions have given rise to *in bello* questions due to their dire economic impact on innocent civilians. Then we document how such sanctions in the ongoing U.S.-Venezuelan episode set the stage for a "responsibility to restore" Venezuela's economy. We

George A. Lopez and Beatrix Geaghan-Breiner, *Establishing an Ethics of Post-Sanctions Peacebuilding* In: *Wicked Problems*. Edited by: Austin Choi-Fitzpatrick, Douglas Irvin-Erickson, and Ernesto Verdeja, Oxford University Press.
© Oxford University Press 2022. DOI: 10.1093/oso/9780197632819.003.0012

conclude with suggested practitioner and scholarly tasks that a "responsibility to restore" necessitates if it is to become a new ethics of action.

How Sanctions Moved from Measured Coercion to Economic War

The conventional argument for economic sanctions has been that such embargos are a reasonable policy choice between weak diplomatic condemnation and going to war against norm-violating nations. Commencing with the Iraqi invasion of Kuwait in August 1990, the United Nations engaged in a decade of sanctions on one dozen nations and groups in order to reverse invasions, stifle civil war, constrain human rights abuses, defund terrorism, and control nuclear proliferation. In the thirty years since the Iraq sanctions, multiple forms of "targeted" or "smart" sanctions have developed to end the indiscriminate impact on civilians and limit societal humanitarian crises. This presumed "magic bullet" aimed harm at leaders rather than entire economies by imposing precision measures on the more narrow workings of an economy. But as elongated smart sanctions episodes unfolded and intensive Western implementation of sectoral and financial sanctions became disastrous for national economies, the moral superiority of "smart sanctions" was debunked. The Trump administration's use of maximum pressure sanctions against North Korea, Iran, and Venezuela, devoid of any substantive diplomacy to achieve their stated political goals, solidified wide acceptance that sanctions constituted economic war.[2]

With thirty years and dozens of cases in mind, we can posit some generalizations regarding which particular sanctions techniques and impacts tend to reveal patterns of inhumane and indiscriminate outcomes that raise *in bello* ethics questions. These demonstrate the need for social and economic restoration post sanctions. They include:

1. The wide application of targeted financial sanctions, including asset freezes of government holdings, personal assets of individuals, and designating banking and financial institutions as "blacklisted" entities that restricts their otherwise legitimate international commerce or finance.
2. Imposing a commodity trade embargo directly, or less directly via restrictions on insurance, taxation, and shipping activity, thus reducing

dramatically the trade revenue of the state. For nations like Iraq, Iran, and Venezuela, where the export of oil is the dominant revenue producer, such commodity sanctions become crippling.

3. As a result of these sanctions, a humanitarian crisis develops as a dramatic decline in key economic indicators then leads to a similar deterioration of social services, especially healthcare, food availability, and water quality. The reaction to appeals for international aid to remedy these problems, and whether humanitarian agencies or goods to provide relief to innocent civilians will be permitted by sanction imposers, emerge as moral issues.

4. To mitigate the worst impacts of sanctions, leaders of a targeted state cultivate sanctions-busting enterprises operated by regional and international criminal networks engaged in illicit financial operations, money laundering, and trafficking in various contraband. The more damaging the sanctions to the national economy, the more criminalized it becomes.[3]

Focusing the Argument: Singular Responsibility for Sanctions' Crippling Impact

Although most analysts accept that sanctions inevitably violate the norm "Do no harm," they have been unengaged regarding how targeted sanctions on a central bank, for example, become such clear *in bello* violations that their impact requires restorative action post-sanctions. We argue this need with a brief analysis where reliable data establish a prima facie claim that the sanctions imposed by a single nation directly induced economic and social infrastructure deterioration. Our analysis is strengthened because the data show that even if all sanctions penalizing the nation were ended immediately, the target's chances of recovery within a reasonable time are very low. Finally, we examine how the issue of humanitarian crises related to sanctions impact is treated by the imposing nation.

In the ongoing U.S. sanctions against the government of Nicolás Maduro in Venezuela, we see the most glaring contemporary example where each of these conditions occur. We describe what the legal community would consider the *corpora delicti* that rise to the level of *in bello* violations inflicted on Venezuelan citizens. We use these to posit a level of "restoration" that *jus post bellum* requires.

In March 2015, President Obama imposed sanctions on Venezuela that included freezing government assets while also imposing a travel ban and asset freeze on key officials in the Maduro government. The goal was to punish the government for regime corruption and its human rights abuses amid internal discord, which the United States claimed posed "an unusual and extraordinary threat to the national security and foreign policy of the United States."[4] The Trump administration, with the aim of toppling the Maduro regime, imposed a maximum pressure campaign of multiple financial sanctions and comprehensive industrial and trade restrictions on most sectors of the Venezuelan economy.

In August 2017, President Trump cut off Venezuela's access to the U.S. financial system, prohibiting it from borrowing in U.S. financial markets. One of the more devastating measures was the U.S. ban on international transactions with the Venezuelan Central Bank in April 2019, which was reinforced by secondary sanctions on any sovereign nation that attempted to do business with Venezuela. By January 2019, the complete assets of the Venezuelan state oil company Petróleos de Venezuela, S.A. (PdVSA), parent of the U.S.-based oil company CITGO, were frozen, and the revenue generated by any purchases of Venezuelan oil by U.S. entities was deposited in U.S. frozen bank accounts. This measure blocked $7 billion in Venezuelan oil assets and resulted in over $11 billion in lost assets over the following year.[5] Since revenue from PdVSA finances almost all public services in Venezuela, the freezing of its assets severely limited the capacity of the government to provide for the basic needs of its citizens. These dramatic shortfalls, combined with Venezuela's total isolation from the global financial system, devastated the country's social and humanitarian infrastructure, creating a major crisis in the health, food, and pure water sectors of the economy. We deal only with the health sector here.

By most estimates, the adequacy of the Venezuelan health infrastructure in 2015 was fully dependent on oil export revenue for meeting a substantial amount of civilian needs. Most concerning, then, as sanctions were imposed, was acquiring life-sustaining medicine and medical supplies. Under U.S. sanctions, foreign banks shut down Venezuela's government accounts and blocked its financial transactions and leverage in the futures market for purchases of critically needed medicines and supplies. In July 2017, Citibank closed accounts belonging to the Central Bank of Venezuela and refused to complete a transaction to procure more than 300,000 doses of insulin when 3 million citizens were insulin dependent.

In November 2017, acting in accordance with rules postulated by the U.S. Treasury Office of Foreign Assets Control, the Belgian financial services company Euroclear retained $1.6 billion that the Venezuelan government had paid to purchase medicine and food for its citizens. The same month, Venezuela paid for a shipment of antimalarial medicine from the Colombian medical laboratory BSN, only to have the U.S.-aligned government in Colombia take action to block the delivery. Similarly, Spain blocked the shipment of 200,000 units of medicine used to treat chronic illnesses, such as diabetes and hypertension.[6]

The linkage between targeted financial sanctions (our category 1) and the economic deterioration of the nation due to sanctions on the nation's primary revenue-producing export (our category 2) is clear. The overall collapse of the Venezuelan economy has been extreme by any standard. GDP per capita (USD) in 2015 was $10,568 and by 2018, $3,411. Inflation has been astronomical, with a 13.3% increase in March 2020 being the lowest monthly increase since June 2017. It is well recognized that such economic collapse can also be attributed to mismanagement by the Maduro government, corruption within the economy, and the global decline in oil prices. But sanctions clearly added to the devastation in some demonstrable, even if partial ways.[7]

As we have illustrated, financial sanctions led directly to a cascading medicine shortage and medical technology crisis, having a dramatic negative impact on the healthcare available to average, innocent civilians (category 3). From 2014 to 2019, the National Survey of Hospitals examined conditions in public and private health facilities spanning the twenty-three states of Venezuela and found that scarcity worsened with each succeeding year of sanctions. In 2014, 55% of the surveyed health units reported shortages in medicine; by 2018, 88% of health units had experienced such shortages.[8]

By any ethical standard, these sanction impacts must be judged as indiscriminate and disproportionate to any harm the Maduro government or Venezuelan citizens had done or threatened to do to the United States or even any political damage the sanctions were imposing on the government. Thus, these outcomes fit the general conditions of violating *jus in bello* norms, especially because the humanitarian consequences of U.S. sanctions are neither collateral nor unanticipated damage. Rather, the explicit logic of maximum pressure sanctions is to inflict such extensive economic and societal pain on civilians that they accomplish the U.S. political goal of deposing Maduro. This became crystal clear when at the onset of the seven-day blackout in

Venezuela on March 7, 2019, Secretary of State Mike Pompeo tweeted, "No food. No medicine. Now, no power. Next, no Maduro."[9] That same month, Pompeo announced at a press briefing that he was "very confident" about the momentum behind the U.S.-led regime change effort: "The circle is tightening. The humanitarian crisis is increasing by the hour. . . . You can see the increasing pain . . . that the Venezuelan people are suffering from."[10]

Remedies in the "Responsibility to Restore"

Various analysts have argued in more robust studies than our brief survey that the *in bello* offenses of most sanctions' episodes demand an immediate end to the brutal sanctions. However appropriate that ethical position, it assumes that the moral obligation to the innocent civilians and the economic demise caused by sanctions subsides with their termination. We maintain that both economic and ethical logic require a "responsibility to restore" key social and economic sectors of a nation in a post-sanctions situation.

An ethic of the responsibility to restore needs to persuade and motivate the UN Security Council and national governments whose policy aims have been achieved via sanctions to engage in a costly process of reconstructing an economy they have, by design, crippled. The Security Council could find persuasive principles supporting restoration in the Universal Declaration of Human Rights that established for every individual "the right to a standard of living adequate for the health and well-being of [one and one's family], including food, clothing, housing, and medical care and necessary social services."[11] In addition, Article 32 of the Charter of Economic Rights and Duties of States, adopted by the UN in 1974, establishes that "no state may use or encourage the use of economic, political, or any other type of measures to coerce another State in order to obtain from it the subordination of the exercise of its sovereign rights."[12] Based on the latter, the Security Council can call on its members to mitigate the ongoing damage of sanctions. But the Council would be unlikely to pass a resolution that obligates a member state to compensate a nation after their termination. This would be the work of regional or international courts.

But neither our highest normative principles nor international law compels the powerful leaders who impose draconian sanctions to compensate recent targets for damages incurred during the episode. What may empower new, restorative ethics and actions are the same general principles that

Alexandra Gheciu and Jennifer Welsh articulated more than a decade ago for why leaders support the imperative to rebuild. Leaders with a sense of special responsibility in the transformational *post-bellum/sanctions* moment or, with a commitment to rebuild self-determination to a nation, are most likely to argue to their domestic political base the need for restoration.[13]

Whatever rationale leads to restoring a besieged economy, effective social and economic remedies are readily available through which sanctions imposers might fulfill their obligation to restore. We state these generally and provide appropriate examples from the U.S.-Venezuela episode:

1. A prompt, facilitated delisting of entities—banks, financial institutions, companies—and individuals placed under sanctions for political and related noncriminal reasons, permitting them to operate freely in the regional and global economy. This also would entail the timely release of frozen assets and other financial resources of the nation and entities.

As of spring 2020, U.S. sanctions have retained and confiscated more than $4.8 billion in assets belonging to the Venezuelan government,[14] including $1.2 billion in gold deposits that were seized by the Bank of England.[15] In addition to the immediate return of these assets to the Venezuelan Treasury, there should be swift compensation for all CITGO revenues that were diverted to U.S.-backed opposition leaders. This money, regardless of who leads the Venezuelan government, would begin to fund its economic recovery and eventually repay the nation's substantial foreign debts. The latter is key to acquiring international development loans.

2. An aggressive, well-financed commitment to warm the "chilly climate" on national banks and financial institutions, thus removing the currency straightjacket that results from sustained, punishing financial sanctions.

This economic pattern enhances the ethical dilemma: banks, development agencies, and international companies are overcompliant with sanctions when they are in place and averse to engaging with a recently sanctioned nation unless the imposers of sanctions inject optimism into the banking and commercial sectors through their own new investments and partnerships. The responsibility to restore must be operationalized in policy and practice by those imposing the sanctions originally.

3. Addressing the socioeconomic welfare costs endured by the civilian population during the sanctions. With the assistance of international agencies, interventions need to reverse health, food, water, and housing crises that may have developed as a result of sanctions. Research studies show that sanctions result in greater inequality, with the heaviest burden hitting both the working class and the poorest within the civil society. Thus the major area for immediate restoration would be to establish the viability of the medical, food, and water sectors of the country.

4. Empower the government to generate a national development plan that assists local economic viability to at least the performance level prior to sanctions. This must include, as noted in (2), remedies to the economic isolation that encumbers the banking and financial sectors, with particular attention to their ability to facilitate currency and other financial movements regionally and globally. This must lead to restoring the capacity of local, business, and financial leaders to wholeness economically regarding their assets and legal liabilities that may have jeopardized them due to their managerial position in a national (non-war-making) enterprise.

5. Plan, participate in, and provide resources necessary to decriminalize the economy and end the corruption that sustains it within various political, legal, and economic institutions. Of all the remedies described briefly here, this is the most substantial commitment required of imposers. It involves a labyrinth of actions in building a rule of law capacity that is stronger than the corruption and criminalization that might be dominating the once sanctioned economy. There are very few models for this task. However, the nature of the problem will demand attention not only to eliminating economic ties to regional and international criminal networks but also to strengthening rule of law systems that may have been eliminated by governing elites.[16]

Future Research and Practice

We have presented here an initial set of arguments for a responsibility to restore post-sanctions environments, bolstered briefly and selectively by the Venezuelan case of U.S.-imposed sanctions damage to innocent civilians. We did not add many dimensions of the U.S.-Venezuela sanctions episode, including judgment on failings of the Venezuelan regime in meeting citizen needs. But we presented clearly the damage sanctions inflict on the health and welfare of innocent civilians within a sanctioned nation and provide ways such damage could be ameliorated. Thus, we pose the ethical dilemma that has emerged in every sanctions case of the past decade: that any nation or international agency that chooses to achieve political objectives through economic sanctions that damage innocent civilians should incur an obligation to restore major segments of the economy that would permit a return to a normal way of life pre-sanctions. What impedes this prospect is the absence of an ethics of the responsibility to restore.

We would like to conclude with a brief overview of some future research opportunities, as well as some practical actions implied by this new ethical principle. We believe it is possible to move the international community forward in solving this dilemma.

First, international legal scholars and ethicists must debate the appropriate term for making societies whole in a post-sanctions situation. We have derived the concept "responsibility to restore" from the humanitarian intervention concept "responsibility to protect" and from peacebuilding experience the concept "responsibility to rebuild" as applicable to the post-sanctions environment. Some will suggest that the rationale and actions required, and the scope of economic resources involved, fit more closely the term "reparations." This claim should initiate an extensive debate that will test whether responsibility to restore can garner a reasonable degree of international consensus and legitimacy.

Second, even though we believe there are strong arguments, and even incentives, for a national sanctioning power to take on restorative actions after sanctions termination, the UN and various other international agencies have substantial roles to play in making sanctioned states economically and politically viable. In particular, these multilaterals are often able to generate codes of conduct in such situations and increase the accountability of nations that pledge restoration. An interesting prospect in this regard should be to mobilize UN agencies, such as panels of experts, the Financial Action Task Force, and the UN Office of Drugs and Crime, that provide information regarding sanctions evasion to warm the chilly climate on banks and advise the decriminalization of an economy.

Third, we are painfully aware that many of our remedies for sanctions and responsibilities for repairing their damages discussed in this chapter open our arguments to some of the same potent critiques of the postwar rebuilding and peacebuilding ventures of the past two decades. An ethics of post-sanctions restoration carries with it the seeds of neocolonialism under the auspices of sanctions recovery. Thus, the task ahead is to model systems of restoration that create and sustain self-determination free of crime, corruption, and external dominance, whereby a nation can thrive through economic inter- and independence.

Finally, long-term ethics turned into the practice of restorative post-sanctions might actually impact the decisional moment, as when an aggressive and punitive set of sanctions is being contemplated by national leaders. Such recognition of this new responsibility to protect from harm and restore would result in future sanctions being designed with a working humanitarian safeguards system. And it might even lead to an ethic whereby sanctions simply are considered neither ethical nor viable as a foreign policy choice due to the harm they inflict on innocent people.

Notes

1. The strongest critique of liberal peacebuilding came from Severine Autesserre, *Peaceland: Conflict Resolution and the Everyday Politics of International Intervention* (New York: Cambridge University Press, 2014). For ethics, see Brian Orend, "Justice after War," *Ethics & International Affairs* 16, no. 1 (March 2002): 43–56. See also Alex J. Bellamy, "The Responsibilities of Victory: *Jus Post Bellum* and the Just War," *Review of International Studies* 34, no. 4 (October 2008): 601–625.

2. For the most telling critiques of smart sanctions and sanctions having little moral credibility, see David Cortright and George A. Lopez, "Are Sanctions Just? The Problematic Case of Iraq," *Journal of International Affairs* 52, no. 2 (Spring 1999): 735–755; Joy Gordon, "A Peaceful, Silent, Deadly Remedy: The Ethics of Economic Sanctions," *Ethics & International Affairs* 13, no. 1 (April 2006): 123–142; Joy Gordon, "Smart Sanctions Revisited," *Ethics & International Affairs* 25, no. 3 (Fall 2011): 315–335; Daniel W. Drezner, "Sanctions Sometimes Smart: Targeted Sanctions in Theory and Practice," *International Studies Review* 13, no. 1 (March 2011): 96–108; Bryan R. Early and Marcus Schulzke, "Still Unjust in Different Ways: How Targeted Sanctions Fall Short of Just War Theory's Principles," *International Studies Review* 21, no. 1 (April 2018): 123–142; and the recent contributions assembled by Joy Gordon, "Roundtable: Economic Sanctions and Their Consequences," *Ethics & International Affairs* 33, no. 3 (September 2019): 275–302.

3. Crime and corruption impacts are understudied in sanctions research, but very good analysis is provided in Peter Andreas, "Criminalizing Consequences of

Sanctions: Embargo Busting and Its Legacy," *International Studies Quarterly* 49, no. 2 (June 2005): 335–360; Bryan R. Early, *Busted Sanctions: Explaining Why Economic Sanctions Fail* (Stanford, CA: Stanford University Press, 2015).

4. U.S. White House Office of the Press Secretary, "*Executive Order: Blocking Property and Suspending Entry of Certain Persons Contributing to the Situation in Venezuela.*" Washington, D.C., 2015. https://obamawhitehouse.archives.gov/the-press-office/ 2015/03/09/executive-order-blocking-property-and-suspending-entry-certain-pers ons-c.

5. U.S. White House Office of the Press Secretary, "White House Daily Briefing," Washington, D.C., January 28, 2019.

6. See Alfred de Zayas, "Report of the Independent Expert on the Promotion of a Democratic and Equitable International Order on His Country Visit to the Bolivarian Republic of Venezuela: Comments by the State." *United Nations Human Rights Council*, September 10, 2018; Unidad Debates Económicos, "The Economic Consequences of the Boycott against Venezuela," *Centro Estratégico Latinoamericano de Geopolítica,* March 12, 2019, https://www.celag.org/economic-consequences-boyc ott-against-venezuela/.

7. Data from Focus Economics, "Venezuela Economic Outlook," in *Focus Economics: Economic Forecasts from the World's Leading Economists*, InSight Crime, July 14, 2020. https://www.focus-economics.com/countries/venezuela

8. Médicos por la Salud and República Bolivariana de Venezuela Asamblea Nacional, "Encuesta Nacional de Hospitales 2018," March 2019. https://cifrasonlinecomve.files. wordpress.com/2018/03/enh-final_2018fin.pdf

9. Secretary Michael Pompeo, Twitter, March 7, 2019, 10:18 p.m., https://twitter.com/ SecPompeo/status/1103872530450771968.

10. U.S. Embassy in Venezuela, "Secretary Pompeo Remarks to the Press on Venezuela," March 11, 2019.

11. UN General Assembly, *Universal Declaration of Human Rights*, December 10, 1948, 217 A (III). https://www.un.org/en/about-us/universal-declaration-of-human-rights

12. UN General Assembly, *Charter of Economic Rights and Duties of States: Resolution/ Adopted by the General Assembly,* December 17, 1984, A/RES/39/163. Refworld | Charter of Economic Rights and Duties of States : resolution / adopted by the General Assembly

13. Alexandra Gheciu and Jennifer Welsh, "The Imperative to Rebuild: Assessing the Normative Case for Postconflict Reconstruction," *Ethics & International Affairs* 23, no. 2 (Summer 2009): 121–146.

14. SURES: Estudios y Defensa en Derechos Humanos, "Unilateral Coercive Measures and Their Impact on the Human Rights of the Venezuelan People" (Caracas, August 2018).

15. Patricia Laya, Ethan Bronner, and Tim Ross, "Maduro Stymied in Bid to Pull $1.2 Billion of Gold from U.K.," *Bloomberg News*, January 25, 2019, https://www.bloomb erg.com/news/articles/2019-01-25/u-k-said-to-deny-maduro-s-bid-to-pull-1-2-bill ion-of-gold.

16. See a synthetic analysis by Fiona Mangan, "Illicit Drug Trafficking and Use in Libya: Highs and Lows," *Peaceworks,* May 2020, https://www.usip.org/publications/2020/05/illicit-drug-trafficking-and-use-libya-highs-and-lows. And see George A. Lopez, "Targeted Sanctions," in *Combating Criminalized Power Structures: A Toolkit*, ed. Michael Dziedzic (Lanham, MD: Rowman & Littlefield, 2016), 61–70.

12

Threading the Needle

Ethical Dilemmas in Preventing Mass Atrocities

Ernesto Verdeja

A central question for human rights practitioners is relatively straightforward: *How can we prevent, or least mitigate, mass atrocities?* But buried within this question are a host of difficult ethical dilemmas involving hard decisions about how to balance various goals—the immediate cessation of killings, the protection of victims, long-term accountability, and the like—that, because of circumstances, are often in deep tension with one another. In exploring these dilemmas here, I draw on existing research, ongoing interviews and conversations with a wide range of prevention practitioners in the Global North and South, and my own occasional consulting work on genocide and atrocity prevention for various governments and nongovernmental organizations (NGOs).

Atrocity prevention remains a pressing human rights issue. Consider just one example: In 2007, Kenya held presidential elections that rapidly devolved into widespread political terror, with around 1,100 civilians killed, hundreds of thousands forced to flee, and massive property damage across the country. International and domestic observers feared "another Rwandan genocide." The historical politicization of identity of the Luo, Kikuyu, and Kalenjin ethnic groups had become a major fault line in the highly contested 2007 struggle for power. Following the election, an investigation by the International Criminal Court found pervasive evidence of crimes against humanity, and an African Union mediation effort identified a host of institutional weaknesses that likely contributed to the violence, including corruption and political interference in the electoral system, the security forces, and the courts. In preparation for the 2013 elections, the government, civil society groups, and international actors launched a concerted atrocity prevention effort, with a focus on reforming the electoral system, the security apparatus, and the judiciary, and also combating widespread hate speech.

Ernesto Verdeja, *Threading the Needle* In: *Wicked Problems*. Edited by: Austin Choi-Fitzpatrick, Douglas Irvin-Erickson, and Ernesto Verdeja, Oxford University Press. © Oxford University Press 2022.
DOI: 10.1093/oso/9780197632819.003.0013

The political alliance of the leaders of two major groups with a history of ethnic violence was also key in reducing the likelihood of further instability.

The 2013 elections were by most accounts a relatively successful, if imperfect, application of atrocity prevention. Kenyans were able to go to the polls without pervasive fear of violence or retaliation. Other prevention cases have been notably less successful, such as the failure to stop the brutal conflicts in eastern Democratic Republic of the Congo, the ethnic cleansing and genocide of Myanmar's Rohingya population, and the failure to generate global pressure to end the ongoing violence in Cameroon. And as Kenya and these other cases show, there remain numerous profound ethical dilemmas at the heart of atrocity prevention work.

In this chapter, I examine what we mean by ethical dilemmas, or "wicked problems," in atrocity prevention, identify some of the most common dilemmas and challenges in current practice, and present some brief thoughts on how to respond to these complex questions. I take a step back from specific cases and reflect on these bigger normative and practical issues in prevention.

International Atrocity Prevention

Atrocity prevention is a catch-all term for a broad set of strategies, practices, and tools to forecast, prevent, mitigate, and stop the reoccurrence of mass killings and other large-scale, sustained, and grievous human rights violations.[1] Prevention has grown significantly since the Rwandan and Bosnian genocides and today consists of an international network of actors that includes intergovernmental organizations like the UN, European Union, and African Union, an enormous number of NGOs and social movements across the globe, and a mix of powerful Western countries and increasingly influential governments from the Global South.[2]

Contemporary prevention practitioners face numerous ethical dilemmas, or situations where a decision is required from among undesirable or problematic alternatives, that is, where there are no obviously superior choices available. Dilemmas are an unavoidable part of atrocity prevention. Indeed, while prevention is animated by ethical convictions, it is also a kind of political activity—it concerns navigating and responding to the demands of powerful political actors in often unstable, dangerous, and rapidly changing contexts, while remaining focused on reducing if not eliminating severe human suffering. Because of this, the most successful practitioners are often effective tacticians and strategists who are capable of adapting to unanticipated contingencies that are part and parcel of political life, while attempting to do so in ways that advance their aims.

For prevention practitioners, a dilemma normally arises when the available set of actions must be justified according to certain ethical principles that cannot be wholly realized or are in tension with one another. To untangle this Gordian knot, I follow several steps. First, I outline the basic ethical principles underpinning mass atrocity prevention, which provide prevention with its overall normative grounding. I then introduce some of the key dilemmas in prevention work, and conclude with some thoughts on how to minimize the severity of dilemmas, even if many of them cannot be eliminated.

Atrocity Prevention Ethics

Ethical Principles

A principle is essentially a basic norm that shapes and informs practical rules of action. Mass atrocity prevention does not have an explicit set of ethical principles, as does humanitarianism.[3] However, we can identity four central principles that underpin prevention: dignity, empathy, universal responsibility, and "Do no harm." Each of these can be grounded in various religious, secular humanist, and cultural traditions, and while important foundational assumptions and details may vary across traditions, the general parameters and substance of each principle are relatively evident and widely endorsed in the practitioner community. I do not have the space to explore their various normative justifications at length, so I present them in summary form.

Dignity refers to the inherent moral worth of all people by virtue of being human (rather than, say, by membership in a particular political, ethnic, religious, or other group). To the extent that all people have dignity, they have commensurate claims to legally binding rights; rights can be understood as legal protections that maintain the integrity of human dignity.

Empathy concerns a general awareness and sensitivity to another's suffering, regardless of their particular identity. It is a morally inflected affective connection to others. Insofar as we can imagine ourselves in the place of those suffering, we hold an empathetic concern for them; to put it slightly differently, our empathy is roused when we witness profound violations of others' dignity.

Universal responsibility involves a moral obligation to care for others in dire need. Responsibility does not end at one's national borders but extends to all people who are in jeopardy of significant human rights violations. The responsibility threshold is normally rather high in atrocity prevention and is met when significant and/or widespread violations are occurring or are increasingly likely to occur. Nevertheless, what responsibility entails practically is heavily debated and involves specifying appropriate responsible agents, the range of actions available to discharge responsibility, how to distinguish between supererogatory and obligatory actions, and how to balance motives and efficacy, among other issues.

The fourth key principle is *Do no harm* (DNH), which has received its most sophisticated treatment in humanitarianism.[4] In prevention, DNH mandates that practitioners should act in such a way that avoids or reduces the risk of exposing people to greater harm. Practitioners should evaluate how various strategies impact conflict dynamics; well-meaning motives or intentions are not enough to adduce ethical value. It is certainly possible to interpret DNH less as a substantive principle and more as a normative "side constraint," that is, laying out the parameters of ethically permissible but not determinative action for adjudicating between competing choices that satisfy DNH. In any case, in actual practice none of the available options may satisfy DNH, so the dilemma becomes how to inflict as little harm as possible rather than cause no harm (this, in turn, assumes qualitatively different harms can be compared).

These four ethical principles justify and orient prevention work and provide much of its normative coherence.

Ethical Ends

Atrocity prevention is about protecting civilians from grave human rights violations, often in the context of significant ongoing political violence. In this, it is of course in line with the general thrust of human rights practice since at least the end of World War II. Nevertheless, its development since the 1990s has distinguished it from other areas of human rights practice, such as conflict prevention and mediation, which are primarily directed toward preventing armed conflict or ending it through a structured, impartial process of negotiation between armed actors (see Laurie Nathan's chapter on mediation in this volume). The boundaries between these various areas are unsurprisingly often blurred, given that all prevention practitioners ultimately seek the reduction and cessation of rights violations. This notwithstanding, for atrocity prevention the absolute legal (and ethical) prohibition on atrocities means that the protection of victims is the primary end, and actions should be oriented around this goal. In conflict prevention, the primary immediate objective is to prevent (or end) armed conflict.

Ethical Means and Effectiveness

What about means and effectiveness? Effectiveness is, on its face, value neutral—it concerns whether actions produce desired results regardless of the latter's normative status. Effectiveness measures success instrumentally with only a secondary concern for guiding principles, motives, or ends. And yet an ethical goal without a practical understanding of how to achieve it is at best inert and at worst risks being unethical, for it may draw attention and resources away from more attainable if limited successes. Effectiveness, then, is itself part of an ethical response.

The instrumental valorization of actions and policies is crucial for prevention practice, but instrumental evaluation should ultimately be grounded (and constrained) by the overarching principles and ends of prevention sketched earlier. Nevertheless, even if means are anchored in ethical principles, their effectiveness is always shaped by the various power relations at play in a given situation.

Ethics and Power

The question of power is central to ethical dilemmas on prevention, but it is rarely theorized in a systematic fashion. Certainly practitioners are highly sensitive to power relations; they know, for instance, that the reason it is so difficult to end the Yemeni Civil War is because of the outsized impact of influential intervening nations that privilege their own strategic interests over human rights. But here I want to foreground power as a central element of ethical dilemmas. To be more precise: by power, I mean the set of material, symbolic, and discursive resources that actors use to advance their interests and goals. These resources are not distributed evenly, and the asymmetries have an enormous impact on all dimensions of prevention.

Asymmetric power relations shape not only the particular choices available to practitioners; they shape the very field of atrocity prevention. These relations frame *what qualifies as an atrocity* worthy of attention, *who is authorized to speak and decide* on a course of action (including who qualifies as a peacebuilder or expert), and, in turn, *who is excluded*. In other words, power is both constitutive and regulative: it both constitutes and regulates the prevention field, and this means it also frames the process of ethical decision-making.

Ethical dilemmas emerge not only because of constraints and uncertainty around specific prevention strategies and tools—on the use of military force, amnesties, and so forth—though these are of serious and pressing importance. Some of these dilemmas, as I will discuss, concern the very structure of contemporary international atrocity prevention architecture, where expert voices from the Global North may marginalize voices from the South. The larger point is that ethical dilemmas emerge precisely because of these uneven topographies or relations of power. In short, there is no Archimedean point outside of messy power relations.[5]

Ethical Dilemmas

There are at least two general clusters of dilemmas in prevention. Dilemmas in *framing* deal with the challenges around defining when atrocities are occurring: what cases qualify as mass atrocities, and who gets to label these. Dilemmas in *acting* concern how to choose from a bad set of options, whether because of problematic consequences or uncertainty about outcomes, and who gets to act "legitimately."

Dilemmas in Framing

The first dilemma is about *selectivity*. Practitioners are faced with the difficult task of generating sufficient public and elite awareness of atrocities to create the necessary political will for preventing, stopping, or at the very least minimizing violations. But how to do this? One way is to compare the ongoing violence to a widely resonant example of evil, such as the Holocaust or the Rwandan genocide, which are commonly invoked to shock collective conscience into action before it is "too late." (Think of how the mantra "Never again" is widely employed.) This may be advisable for breaking public apathy, but it also risks ignoring those cases that do not fall into straightforward narratives about violence. The dilemma is that in raising some cases as demanding our attention, others are sidelined. Bringing attention to some cases risks privileging them over others. For every next possible Rwanda, there are many unrecognized Cameroons.

A related but distinct dilemma is the *simplification* of mass atrocity narratives to catalyze public awareness, which in turn may complicate or hamper responses or lead to public attention fatigue when easy solutions aren't forthcoming. This was a major concern, for instance, with the reductive ways in which the genocide and civil war in Darfur, Sudan, were publicly presented by some human rights groups, or the fiasco around the Kony 2012 media campaign to explain, in highly reductive terms, the conflict in northern Uganda. The dilemma is made more acute when substantial popular attention—normally a good thing—distorts prevention priorities and policies.

A final framing dilemma is about who is recognized as a legitimate interlocutor—*who gets to speak*. The international atrocity prevention community privileges technical sophistication and expert knowledge,[6] and many Global North actors have a diverse and increasingly sophisticated set of prevention tools, strategies, and "lessons learned" that are applied across wider sets of sometimes disparate cases. The uptake is that these actors have the material resources, symbolic status, and discursive facility to shape discussions about how atrocities are framed, while local actors in the Global South are often ignored or only superficially consulted, a point underscored by Reina Neufeldt's and Phil Gamaghelyan's respective contributions to this book. The dilemma, then, concerns how to avoid reinforcing a kind of "neocolonial"[7] influence over responses to Global South crises and instead leverage the extensive resources, networks, and influence of the Global North while ensuring greater input and decision-making influence of regional and local actors.

Dilemmas in Acting

A second cluster of dilemmas concerns action. These can take many forms, such as how to insulate an NGO from donor pressure,[8] but for space reasons I will focus on only three of the major dilemmas that consistently arise in atrocity prevention.

The first involves the pitfalls of negotiation, or *acceding to the demands* of violent actors. Often, the only way to end violence, even if temporarily, is to provide perpetrators credible assurances of security or formal recognition. They may demand guarantees of protection (amnesty, continued political or military control, etc.) as a condition for putting away their weapons. Formulating acceptable trade-offs requires a clear articulation of primary goals: is it to stop or limit current atrocities, allow vulnerable populations to escape safely, prevent future outbreaks of violence, or something else? If there are multiple objectives, how should success be measured? The dilemma may be compounded by temporal valorization: if protecting civilians in the present necessitates security guarantees for violators, this may still leave conditions for future violence, especially when the deep sources of grievances and injustice haven't been addressed.

The use of force presents another dilemma: How and when should a credible threat of substantial retaliatory force be used against perpetrators? The logic is straightforward: a threat is needed to incentivize perpetrators to stop killing, and where that fails, force is employed to end or mitigate ongoing massacres—this underpinned the North Atlantic Treaty Organization interventions in Kosovo and Libya. The dilemma for prevention practitioners stems from having to endorse the use of violence to secure peace, which prima facie risks violating the DNH principle, especially when the likelihood of success is not clear. The UN's Responsibility to Protect (R2P) norm provides some important guidance on the use of force, but the track record of military "humanitarian" interventions is mixed at best, and R2P has been manipulated by countries pursuing their own narrow interests. (Russia, for instance, invoked R2P in its conflicts with Georgia and Ukraine.)

A third dilemma involves the longer term, namely the trade-offs and compromises of *collaborating with states* to create the mechanisms and response architecture to prevent future atrocities domestically, regionally, and globally. The dilemma is especially sharp when working with powerful Western governments (the United States, Great Britain, France, etc.) that are indispensable for advancing prevention but also prioritize their national interests when it suits them, which can make them simultaneously human rights advocates and drivers of violence and instability. A somewhat different set of challenges involves working with governments that are domestically repressive but selectively support prevention elsewhere, especially in their own region (consider, for instance, Rwanda or Uganda). Of course, some states, such as Syria, Myanmar, and their supporters, like Russia and China, show minimal commitment to atrocity prevention, but courting them is necessary at times. The dilemma, then, lies in knowing when and how to work with actors who may be simultaneously key prevention partners while also implicated in conflict. In essence, the dilemma involves how to navigate state hypocrisy and provide states with persuasive incentives to support prevention.

Ethical Responses

These dilemmas are typical of the challenges faced by the atrocity prevention community. They involve how to protect civilians while navigating an uneven terrain of political power that extends over a range of scenarios and time horizons. Furthermore, ethical evaluation and decision-making are themselves partly context-specific; there is no abstract set of rules that can consistently generate a "right" and final answer in messy, rapidly shifting circumstances with multiple actors and imperfect real-time information. Given all of this, it is likely more useful to outline some general guidelines for ethical action that can help lessen the space between the four principles— dignity, empathy, universal responsibility, and DNH—on the one hand and the practical need for effectiveness on the other. I divide these into normative and practical guidelines.

Normative Guidelines

Harmony with normative ends: Actions should be directed toward reducing widespread human suffering, specifically mass atrocities. Policies, programs, and decisions should align with these goals, though varying time horizons may raise challenges about practical sequencing.

Harmony with principles: Actions should conform as much as possible to the basic principles of dignity, empathy, universal responsibility, and refraining from harm. This means ensuring that civilian groups are treated as ends in themselves and not merely instrumentally to further some other goal. It also requires an inclusive understanding of responsibility, one that does not draw moral distinctions between deserving and undeserving civilians.

Partnership: Practitioners should support, to the greatest extent possible, partnering with the most affected populations and local human rights advocates to ensure their agency and autonomy. Civilians are more than passive objects for protection; their interests and goals should be a central part of prevention work.

Reflective practice: It is crucial to avoid offering only a technical response to what are overwhelming ethical challenges. Given how circumstances change, practitioners should adopt a perspective of reflective practice, where decisions and options are subject to sustained scrutiny in terms of normative coherence, levels of involvement of affected groups, and overall practical efficacy. It is important, in other words, to be explicit about one's moral reasoning and justifications for particular strategies and policies. This also means analyzing one's subject position: Who is speaking and who gets to speak, and are traditionally ignored voices part of the conversation?

Practical Guidelines

Practical ends: Clarify the specific aims of particular policies and actions, whether and how they may be in tension, and how they contribute individually and collectively to the overarching end of reducing atrocities. What is success: a return to the status quo prior to the outbreak, or something more transformative and permanent? Identify, if appropriate, a hierarchy of particular aims, and clearly justify their prioritization. Rather than adopt a series of piecemeal approaches to specific problems, provide a clear understanding of how practical ends cohere, which can facilitate the development of integrated responses to the complexity of large-scale harms. Of course, circumstances change, and practitioners must assume that unforeseen events will raise new challenges and require quick and unplanned adjustments. However, a well-developed understanding of how practical ends connect can bring greater coherence to these efforts.

Relevant parties: Identify relevant actors, including sympathetic parties, spoilers, and bystanders, and focus on alliance building and leveraging collective pressure on key players. Identify the full range of targeted civilian groups in need of protection.

Implementation, monitoring, and evaluation: Lay out clearly the required steps and sequencing for implementing preferred policies, and whether they work at cross-purposes. Additionally, clarify the standards for assessing success and failure and how these standards are operationalized, as well as the relevant types of necessary evidence.

Strengthen global, regional, and national *prevention architectures* for future responses. This involves moving beyond short- and mid-term responses to long-term solutions. A robust prevention architecture will not prevent all future crises, of course, but it can lessen their frequency and severity and may reduce the intensity of some dilemmas by replacing ad hoc responses and unplanned crisis management with more sustained and effective mechanisms for prevention. To the extent that atrocity prevention is integrated within a broader human rights and development framework, it can be directed toward targeting the root causes of violence—poverty, inequality, systemic discrimination, and the like—well before atrocities begin. To be sure, greater institutionalization can bring new dilemmas: the UN has an institutional bias toward bureaucratized process over substantive outcomes, which can lead to paralysis when what are needed are rapid, nimble, and meaningful responses, creating a dilemma for practitioners who seek both international legitimacy and swift and robust results.

Preventing mass violence is arduous and challenging, fraught with ethically complicated choices. It requires a steadfast commitment to principles and a belief that the world can be made more just and safer with sufficient work and struggle, but it can also test the resolve of even the most committed activists. But even if dilemmas can never be overcome completely—for many of them are constitutive of the very politics of prevention—they can be mitigated and lessened over time with ethically reflective and informed practice.

Notes

1. Normally, this includes genocide, war crimes, crimes against humanity, and ethnic cleansing. See United Nations, *Framework for Analysis of Atrocity Crimes: A Tool for Prevention* (New York: United Nations, 2014).
2. James Waller, *Confronting Evil: Engaging Our Responsibility to Prevent Genocide* (Oxford: Oxford University Press, 2016).
3. See, for instance, Sphere, Humanitarian Charter, accessed September 17, 2021, https://spherestandards.org/humanitarian-standards/humanitarian-charter/; International Committee of the Red Cross, Code of Conduct for the International Red Cross and Red Crescent Movement and Non-Governmental Organizations (NGOs) in Disaster Relief, December 12, 1994, https://www.icrc.org/en/doc/resources/documents/publication/p1067.htm; International Federation of Red Cross and Red Crescent Societies, The Seven Fundamental Principles of the Red Cross and Red Crescent Societies, https://www.icrc.org/sites/default/files/topic/file_plus_list/4046-the_fundamental_principles_of_the_international_red_cross_and_red_crescent_movement.pdf.

4. Mary B. Anderson, *Do No Harm: How Aid Can Support Peace—or War* (Boulder, CO: Lynne Rienner, 1999); Hugo Slim, *Humanitarian Ethics: A Guide to the Morality of Aid in War and Disaster* (Oxford: Oxford University Press, 2015).

5. Catherine Goetze, *The Distinction of Peace: A Social Analysis of Peacebuilding* (Ann Arbor: University of Michigan Press, 2017); Ernesto Verdeja, "Critical Genocide Studies and Mass Atrocity Prevention," *Genocide Studies and Prevention* 13, no. 3 (2019): 111–127.

6. Verdeja, "Critical Genocide Studies and Mass Atrocity Prevention," 121.

7. Slim, *Humanitarian Ethics*, 10.

8. Peace Direct, *Atrocity Prevention and Peacebuilding*, 2018. https://www.peacedirect.org/us/wp-content/uploads/sites/2/2018/04/Atrocity-Prevention-Report_PD.pdf.

13

Whither the Villains?

The Ethical Dilemma in Armed Conflict

Laurie Nathan

Over the past two decades I've been involved in national and international peace mediation in high-intensity conflicts. This form of conflict resolution throws up profound ethical dilemmas. One of the most serious dilemmas relates to the fate of villains responsible for gross human rights abuses and excessive use of force: Should the villains be included in the mediation and negotiated settlement, thereby compromising justice and accountability? Or should they be excluded from the mediation process and outcome, thereby perpetuating hostilities by ensuring that the villains will fight to the bitter end? The dilemma has no simple solution because it pits the good norms of peace and stability against the good norms of justice and accountability.[1]

In 2005–2006 I participated in the African Union mediation to end the violent conflict in the Darfur region of Sudan.[2] A rebellion had broken out two years earlier in opposition to the government's oppression of Darfuri communities. The government responded to the rebellion with such ferocity that it was accused of genocide. Mediation seemed to be the only way to end the slaughter and protect vulnerable communities. If successful, though, mediation would result in a power-sharing deal between the rebels and the government. It would leave the government in power and let most of the perpetrators of war crimes go unpunished. The International Criminal Court (ICC) had indicted Sudan's president Omar al-Bashir, but this only reinforced the regime's intransigence and reduced the likelihood of ending the conflict peacefully. In these circumstances, what was the ethically appropriate stance for the mediators and the international community: to pursue justice and accountability, or peace and security?

International efforts to end armed conflicts—as in Darfur, Côte d'Ivoire, Libya, South Sudan, Syria, and Yemen—are often wracked by this dilemma.

Laurie Nathan, *Whither the Villains?* In: *Wicked Problems*. Edited by: Austin Choi-Fitzpatrick, Douglas Irvin-Erickson, and Ernesto Verdeja, Oxford University Press. © Oxford University Press 2022. DOI: 10.1093/oso/9780197632819.003.0014

The dilemma arises from the brutal nature of these conflicts. One or more of the conflict parties are likely to have committed gross human rights abuses and engaged in indiscriminate violence. These problems do not only arise during the conflict. The conflict itself may be a response to persistent authoritarianism and repression. The dilemma, then, is whether international peacemaking initiatives should seek to remove the villains from power, which may require external military action, or pursue a mediation process that includes the villains and might lead to some kind of power-sharing deal with them.

From an ethical perspective, mediation in civil wars offers the prize of peace and may exact the price of compromising accountability and justice by allowing the villains to continue to occupy positions of power.

The ethical dilemma and challenges are strongly evident when the international community is divided on the most appropriate strategy for ending a conflict, with some international actors favoring the use of military force to oust the villains while other international actors insist that an inclusive mediation process is preferable. In order to capture these tensions, this chapter first presents an overview of the acrimonious debates over the international response to the conflicts in Côte d'Ivoire (2010–2011), Libya (2011) and Syria (2011–2018). This is followed by an analysis that explains why the dilemma is so difficult to resolve. The dilemma is characterized as a "situational incompatibility of good norms," which arises in situations where a hard choice has to be made between two sets of good norms that are generally compatible with each other.

Côte d'Ivoire, 2010–2011

In late 2010, eight years after the start of the civil war in Côte d'Ivoire, the country held a presidential election that was expected to usher in an era of peace and stability. The election was won by the leader of the opposition, Alassane Ouattara. Far from ending the crisis, however, this outcome reignited violence when President Laurent Gbagbo refused to concede defeat. Ouattara's supporters mounted protests and aligned themselves with rebel forces, and Gbagbo mobilized army units and militia to suppress them. After several months of clashes, the country appeared to be on the brink of returning to war.[3]

The main international actors that sought to prevent another war—the African Union (AU), the Economic Community of West African States (ECOWAS), and the United Nations (UN)—agreed that Ouattara had won a free and fair election. But they had severe disagreements on normative priorities, strategy, and desired outcomes.

At the early stage of the electoral dispute, the AU deployed a series of high-profile mediators.[4] The first was Thabo Mbeki, the former South African president, who advocated a negotiated settlement between Gbagbo and Ouattara. When Mbeki failed to make headway, the AU dispatched Kenyan prime minister Raila Odinga as its mediator. Before leaving Kenya, Odinga threatened force if Gbagbo did not step down. Unsurprisingly, Gbagbo refused even to meet with Odinga. Thereafter an AU mediation panel comprising five heads of state proposed that Ouattara be installed as president, that he set up a government of national unity excluding Gbagbo but including his party, and that a blanket amnesty be conferred.[5] Gbagbo rejected this proposal.

ECOWAS, on the other hand, responded to Gbagbo's refusal to step down by threatening to use force to remove him.[6] As the fighting and atrocities escalated in early 2011, the regional organization appealed to the UN Security Council (UNSC) to bolster the mandate of the UN Operation in Côte d'Ivoire (UNOCI) and ensure a transfer of power to Ouattara.[7] The Security Council obliged, authorizing UNOCI to "use all necessary means" to protect civilians.[8] With military support from the UN and France, Ouattara defeated Gbagbo. Gbagbo was later transferred to The Hague to await trial at the ICC.

From the perspective of those who favored mediation and power-sharing, the highest priority was to avoid a relapse into civil war. As Mbeki put it, a negotiated settlement was the only way to avert "a very destructive war which will inevitably result in many civilians losing their lives and much property destroyed."[9] In the interest of peace, the AU mediation panel proposed a softer compromise that would accommodate Gbagbo's party but not Gbagbo himself.

For the UN, Odinga, and ECOWAS, a power-sharing deal would subvert the democratic election and reward Gbagbo's resort to violence. A UN official argued publicly that Mbeki's promotion of a negotiated settlement undermined democratic tenets and would set a dangerous precedent.[10] Odinga lambasted Gbagbo's "rape of democracy" and called on the AU to "develop teeth instead of sitting and lamenting all the time or risk becoming irrelevant."[11] ECOWAS's opposition to power-sharing was grounded in its concern that tolerance of Gbagbo's stance would lead to similar undemocratic action elsewhere in the region. ECOWAS was also reportedly influenced by Mbeki's mediation of a power-sharing deal in Zimbabwe in 2008, which ECOWAS viewed as a failure because it left President Robert Mugabe in power after he had lost an election.[12]

Libya, 2011

In early 2011 street protests in Libya were mounted against the authoritarian regime of Muammar Qaddafi. The protests were met with state violence and evolved rapidly into a full-scale civil war. The UNSC reacted quickly by passing Resolution 1970, which called for an immediate end to the violence and for "steps to fulfill the legitimate demands of the population."[13] Considering that the government's attacks against the civilian population may amount to crimes against humanity, the UNSC referred the situation to the ICC.[14]

When Qaddafi's forces appeared poised to take the rebel-held town of Benghazi, the UNSC issued Resolution 1973. It demanded an immediate ceasefire and authorized member states to "take all necessary measures" to enforce a no-fly zone and protect civilians and civilian-populated areas under threat of attack. It welcomed the UN Secretary-General's appointment of a mediator for Libya and approved the decision by the AU to play a mediating role. Resolution 1973 thus endorsed both enforcement action and mediation to end the crisis. The enforcement action was undertaken by NATO, whose air strikes helped to destroy Qaddafi's forces and defeat the regime in August 2011.

The objective of the three Western permanent members of the UNSC ("the P3")—France, the United Kingdom, and the United States—was forcible regime change.[15] In light of Qaddafi's history of repression and alleged war crimes in suppressing the rebellion, the P3 were emphatic that his immediate departure was not negotiable.[16] This position was supported by, among others, the European Union, the League of Arab States (LAS), the Gulf Cooperation Council, and the Organization of Islamic States.

At the same time, there was considerable international opposition to the regime change agenda and reliance on external use of force. The opponents included the AU, the five members of the UNSC that abstained from voting on Resolution 1973 (Brazil, China, India, Germany, and Russia), and some of the NATO countries that declined to join the military campaign. These actors argued that the crisis should be resolved through mediated negotiations and not force; that Qaddafi's exit should be the subject of negotiations and not a precondition for negotiations; and that the Security Council's enforcement mandate to protect civilians should not be extended to regime change.

The exclusionary agenda of the P3 stifled the possibility of ending the crisis through mediation. The ICC referral, which was followed in June 2011 by the ICC issuing warrants for the arrest of Qaddafi and one of his sons, made it improbable that Qaddafi would see negotiations as a viable way out of the conflict. The AU complained that the warrants would "pour oil on the fire" rather than help to end the war.[17] The International Crisis Group warned presciently that the demand that Qaddafi not only vacate power but also leave the country and face trial at the ICC "virtually . . . ensure[d] that he [would] stay in Libya to the bitter end and go down fighting."[18]

The exclusionary agenda made it impossible even to negotiate a ceasefire. The P3 and the rebels insisted that Qaddafi's exit was a precondition for a ceasefire, and they therefore rejected ceasefire proposals made by the Qaddafi regime, the AU, and other third parties. Their argument that Qaddafi would never honor his undertaking to cease fighting ignored the fact that in numerous civil wars, skepticism about the conflict parties' professed commitment to a ceasefire has been the subject of negotiations rather than a reason to avoid negotiations.[19]

The overthrow of Qaddafi and the failed transition process thereafter enabled the growth of extremist Islamist groups in Libya and the flow of weapons into the hands of terrorist militias across North Africa, West Africa, and the Middle East. Since the fall of the regime, Libya has been characterized by state failure, armed conflict, and anarchy.[20]

Syria, 2011–2018

For more than a year after the start of the Syrian Civil War in March 2011, divisions among the permanent five members of the UNSC ("the P5") prevented the Council from issuing a resolution on the crisis. The UN General Assembly filled the gap with Resolution 66/253, which requested the UN Secretary-General to appoint an envoy. Following his appointment, Kofi Annan crafted the "six-point plan" that called on the Syrian government to stop the fighting, achieve an effective UN-supervised cessation of violence, and work with the UN envoy on "an inclusive Syrian-led political process to address the legitimate aspirations and concerns of the Syrian people."[21]

As hostilities and fatalities soared, Annan endeavored to forge a united international front. He convened the Action Group for Syria, comprising the UN, the LAS, the EU, the P5, and some of the regional powers involved in the conflict. The Action Group produced the Geneva Communiqué, which called for the formation of a transitional governing body that would oversee the transition to a new constitutional and democratic order.[22] When the UNSC moved to ratify the Communiqué, however, the P3 sought to include enforcement provisions, causing Russia and China to veto the draft resolution. Two weeks after the failed vote, Annan resigned, citing the Syrian government's refusal to implement the six-point plan, the escalating military campaign of the opposition, and the international community's disunity.[23]

At the heart of the international disunity was the question of President Bashar al-Assad's future. Those pushing for his departure as a precondition for the transition were the Syrian opposition groups, the P3, the LAS, and Turkey. The international actor objecting most strongly to this was Russia, but a much larger number of countries shared its view. This was evident in the UN General Assembly resolution, which declined to back the demand for Assad's immediate departure. On the contrary, the resolution called for an inclusive Syrian-led political process that would include a political dialogue between the government and the Syrian opposition.

Annan believed that the appropriate strategy for achieving Assad's exit was not through public proclamations but through pressure from the president's allies, Iran and Russia.[24] For Annan, the issue was when, and not if, Assad would leave, and he maintained that this issue was an unnecessary obstacle to initiating a peace process.[25] The P3's concerted push for Assad's exit impeded the efforts of Annan and other mediators to end the war through mediation: it weakened international support for mediation, reinforced Assad's resistance to negotiations, and strengthened opposition to negotiations from the rebels, who believed that the P3 would take decisive military action against Assad.[26] Annan complained that the P3, through their military support for the rebels, had undermined the UNSC's agreement on the need for a peaceful settlement.[27]

The P3 were not concerned that their position might jeopardize the UN's mediation efforts since they were convinced that Assad, like Qaddafi before him, would be defeated.[28] From a humanitarian perspective, this prognosis was tragically mistaken. Analytically, it illustrates the point that the success or failure of an exclusionary position depends primarily on the balance of power and preponderance of force, which, in the case of Syria, have ultimately favored Assad. By 2018—after seven years of devastating civil war that had destroyed large parts of the country, provoked a humanitarian crisis of epic proportions, and exacerbated regional instability—Assad had effectively achieved military victory.[29]

Ousting the Villains or Sharing Power with Them?

The ethical dilemma in these conflicts can be analyzed in terms of international norms, which are shared standards or expectations of appropriate behavior that apply to and are accepted by states and intergovernmental organizations.[30] Because norms delineate appropriate types of behavior, they necessarily have an ethical and prescriptive character.[31] They evolve over time and are contested as new norms challenge established ones. The international relations literature concentrates on "good" transnational norms that challenge prevailing "bad" norms.[32] The former include, for example, the antislavery and anti-apartheid movements, decolonization, gender equality, respect for human rights, and restrictions on the use of certain weapons.[33]

In the cases of Côte d'Ivoire, Libya, and Syria, the political tension among international actors hinged on the question of whether the conflict should be resolved through forceful measures that removed the villains from power or through inclusive mediated negotiations that might lead to a power-sharing arrangement. Whereas the international actors calling for the villains to be ousted prioritized the norms of democracy, justice, and accountability, those calling for mediated negotiations prioritized the norms of peace and security. The international friction thus reflected an acute ethical dilemma. In contrast to contests between good norms and bad norms, in these cases two sets of good norms were pitted against each other.

Of course there is no inherent or general tension between the norms of democracy, accountability, and justice on the one hand, and the norms of peace and security on the other. Many of the relevant international actors have endorsed both sets of norms (e.g., the AU, ECOWAS, the EU, the P3, and the UN), and the prevailing international discourse regards these norms as complementary and mutually reinforcing.[34] In situations of armed conflict, however, the two sets of good norms may be in tension such that international actors engaged in peacemaking feel compelled to prioritize one set of norms over the other. Accordingly, the ethical dilemma can be understood as a "situational incompatibility of good norms."

The dilemma is not reducible to a choice between the respective *benefits* of the two sets of good norms. It must also weigh the potential political and humanitarian *costs* of favoring one set of norms over the other: Is the prize of peace worth the cost of allowing the villains to share power and escape accountability, or is the determination to oust the villains worth the cost of prolonging the conflict by ensuring that they fight to the bitter end? Both of these costs are morally abhorrent.

The particular dynamics of armed conflict heighten the acuity and complexity of the dilemma, as well as the difficulty of resolving it. The reasons for this relate to strategic uncertainty, unintended consequences, and humanitarian concerns.

Strategic uncertainty arises from the volatility and unpredictability of armed conflict and from the fact that the external actors are not in control of the situation. To the extent that anyone is in control, it is the belligerent parties, locked in mortal combat. External actors therefore cannot be certain that their interventions will achieve the intended objective. In the cases under consideration, there was no guarantee that mediation would have succeeded in brokering a negotiated settlement or that a settlement, if reached, would have been stable and respectful of human rights. Nor were the outcomes of external military action predictable. As it turned out, in Côte d'Ivoire the villain was defeated and relative stability has since prevailed; in Libya the villains were also defeated but the country was plunged into a protracted crisis of violence and state failure; and in Syria the villains have triumphed militarily.

In conditions of strategic uncertainty, the ethical dilemma is exacerbated by the problem that international interventions might not only fail to achieve their objective but might also have unintended harmful consequences. In Libya, for example, the NATO bombing campaign spurred the collapse of the state, created a long-term power vacuum with grave political and security implications for Libya and its neighbors, intensified the refugee emergency, and contributed to the flow of weapons into the hands of terrorist militias across North Africa, West Africa, and the Middle East.[35]

Humanitarian concerns add another layer of complexity and difficulty to the ethical dilemma. The extreme violence and massive human suffering that characterize armed conflict generate intense political and normative pressure on the international community to respond quickly and decisively. The UN and regional organizations cannot weigh up alternative courses of action in a contemplative and patient manner. Notwithstanding the risks, the humanitarian imperative is to take decisive action quickly. Abstaining from all action on the grounds of uncertainty and the principle "Do no harm" is not a viable political option.

Space constraints preclude a more comprehensive examination of the nuances and variations that are typically at play in relation to the conflict parties and to external and domestic strategies in situations of armed conflict: the cases discussed here focused on a particular villain, but in armed conflicts there are bound to be degrees of villainy across the board and there are few, if any, angels; the external strategies are not limited to mediation and use of force but include sanctions, arms embargoes, and suspension from regional bodies; the options for transitional justice are not restricted to trial at the ICC but include a range of domestic mechanisms; and there are many permutations of transitional and longer-term power-sharing.

Notwithstanding these variations and nuances, the central ethical dilemma in international responses to armed conflict hinges on a choice of ousting the villains through military force or including them in mediated negotiations that might lead to an inclusive settlement and power-sharing arrangements. In this chapter I've refrained from offering "solutions" to the dilemma. In the nature of ethical dilemmas, there are no obvious solutions. This particular dilemma is ethically acute and complex because it requires a choice to be made between the good norms of democracy, justice, and accountability on the one hand, and the good norms of peace and security on the other. The resolution of the dilemma is therefore bound to be painful— whatever choice is made by actors in a given conflict will exact an ethical price and often a political and humanitarian price as well. Resolving the dilemma is further complicated by conflict volatility, strategic uncertainty, and grave humanitarian concerns. The most cogent policy advice is that international actors, when confronted by the dilemma in a given case, should proceed with humility. They should avoid the hubristic conviction that they have a definitive solution to the problem before them and that their solution will indubitably lead to a good outcome. The cases discussed in this chapter belie that conviction.

Notes

1. This chapter is an abridged version of Laurie Nathan, "The International Peacemaking Dilemma: Ousting or Including the Villains?," *Swiss Political Science Review* 26, no. 5 (2020): 468–486.

2. Laurie Nathan, "No Ownership, No Peace: The Darfur Peace Agreement," Working Paper 5, Crisis States Research Centre, London School of Economics, 2006.

3. See Alex Bellamy and Paul Williams, "The New Politics of Protection? Côte d'Ivoire, Libya and the Responsibility to Protect," *International Affairs* 87, no. 4 (2011): 825–850.

4. Kasaija Apuuli, "The African Union's Notion of 'African Solutions to African Problems' and the Crises in Côte d'Ivoire (2010–2011) and Libya (2011)," *African Journal of Conflict Resolution* 12, no. 2 (2012): 135–160.

5. African Union, "Report of the High Level Panel of the African Union for the Resolution of the Crisis in Côte d'Ivoire," March 10, 2011.

6. Economic Community of West African States, "Final Communiqué, Extraordinary Session of the Authority of Heads of State and Government on Côte d'Ivoire," press release 192/2010, December 24, 2010.

7. Economic Community of West African States, "Resolution A/RES.1/03/11 of the Authority of Heads of State and Government on the Situation in Côte d'Ivoire," press release 043/2011, March 25, 2011.

8. United Nations Security Council, Resolution 1975/2011, March 30, 2011.

9. "Mbeki: Côte d'Ivoire Rivals Must Talk to End the Crisis," *Mail and Guardian*, January 24, 2011, http://mg.co.za/article/2011-01-24-mbeki-cocircte-divoire-rivals-must-talk-to-end-crisis.

10. Vijay Nambiar, "Dear President Mbeki: The United Nations Helped Save the Ivory Coast," *Foreign Policy*, August 17, 2011, http://foreignpolicy.com/2011/08/17/dear-president-mbeki-the-united-nations-helped-save-the-ivory-coast/.

11. Nicolas Cook, "Côte d'Ivoire's Post-Election Crisis," CRS Research Report for Congress, Congressional Research Service, January 28, 2011, 14, https://www.refwo rld.org/pdfid/4d58e5832.pdf.

12. Colum Lynch, "On Ivory Coast Diplomacy, South Africa Goes Its Own Way," *Foreign Policy*, February 23, 2011, 3, http://foreignpolicy.com/2011/02/23/on-ivory-coast-diplomacy-south-africa-goes-its-own-way/.

13. United Nations Security Council, Resolution 1970/2011, February 26, 2011.

14. See International Crisis Group, "Popular Protest in North Africa and the Middle East (V): Making Sense of Libya," *Middle East/North Africa Report*, no. 107 (2011). https://www.crisisgroup.org/middle-east-north-africa/north-africa/libya/popular-protest-north-africa-and-middle-east-v-making-sense-libya.

15. Micah Zenko, "The Big Lie about the Libyan War," *Foreign Policy*, March 22, 2016, https://foreignpolicy.com/2016/03/22/libya-and-the-myth-of-humanitarian-inter vention/.

16. Allegra Stratton, "Obama, Cameron and Sarkozy: No Let Up in Libya until Gaddafi Departs," *The Guardian*, April 14, 2011, https://www.theguardian.com/world/2011/apr/15/obama-sarkozy-cameron-libya.

17. Franz Wild, "ICC Warrant 'Pours Oil on Fire' in Libya, African Union Says," *Bloomberg*, June 29, 2011, http://www.bloomberg.com/news/2011-06- 29/icc-warrant-pours-oil-on-fire-in-libya-african-union-says-1-.htm.

18. International Crisis Group, "Libya: Achieving a Ceasefire, Moving toward a Legitimate Government," statement, May 13, 2011, https://www.crisisgroup.org/middle-east-north-africa/north-africa/libya/libya-achieving-ceasefire-moving-toward-legitimate-government.

19. International Crisis Group, "Popular Protest in North Africa and the Middle East," 28–29.

20. Lisa Watanabe, "UN Mediation in Libya: Peace Still a Distant Prospect," *CSS Analyses in Security Policy*, no. 246 (2019): 1–4.

21. Reuters Staff, "Text of Annan's Six-Point Peace Plan for Syria," Reuters, April 4, 2012, https://www.reuters.com/article/us-syria-ceasefire/text-of-annans-six-point-peace-plan-for-syria-idUSBRE8330HJ20120404.

22. Action Group for Syria, "Final Communiqué," June 30, 2012, https://peacemaker.un.org/sites/peacemaker.un.org/files/SY_120630_Final%20Communique%20of%20the%20Action%20Group%20for%20Syria.pdf.

23. Kofi Annan, "My Departing Advice on How to Save Syria," *Financial Times*, August 2, 2012, https://www.ft.com/content/b00b6ed4-dbc9-11e1-8d78-00144feab49a#ixzz22PUj6wxm.

24. Tom Hill, "Kofi Annan's Multilateral Strategy of Mediation and the Syrian Crisis: The Future of Peacemaking in a Multipolar World?," *International Negotiation* 20, no. 3 (2015): 463.

25. Hill, "Kofi Annan's Multilateral Strategy of Mediation and the Syrian Crisis," 463.

26. Raymond Hinnebusch and I. William Zartman with Elizabeth Parker-Magyar and Omar Imady, *UN Mediation in the Syrian Crisis: From Kofi Annan to Lakhdar Brahimi* (New York: International Peace Institute, 2016); Hill, "Kofi Annan's Multilateral Strategy of Mediation and the Syrian Crisis."

27. Hill, "Kofi Annan's Multilateral Strategy of Mediation and the Syrian Crisis," 470–471.

28. Hill, "Kofi Annan's Multilateral Strategy of Mediation and the Syrian Crisis," 449, 462–463.

29. Mara Karlin, "After 7 Years of War, Assad Has Won in Syria. What's Next for Washington?," Brookings Institute, February 13, 2018, https://www.brookings.edu/blog/order-from-chaos/2018/02/13/after-7-years-of-war-assad-has-won-in-syria-whats-next-for-washington/.

30. Sanjeev Khagram, James Riker, and Kathryn Sikkink, "From Santiago to Seattle: Transnational Advocacy Groups and Restructuring World Politics," in *Restructuring World Politics: Transnational Social Movements, Networks, and Norms*, ed. Sanjeev Khagram, James Riker, and Kathryn Sikkink (Minneapolis: University of Minnesota Press, 2002), 14.

31. Martha Finnemore and Kathryn Sikkink, "International Norm Dynamics and Political Change," *International Organization* 52, no. 4 (1998): 891–892.

32. Jeffrey Checkel, "The Constructivist Turn in International Relations Theory," *World Politics* 50, no. 2 (1998): 339.

33. See, for example, Audie Klotz, "Transnational Activism and Global Transformations: The Anti-Apartheid and Abolitionist Experiences," *European Journal of International Relations* 8, no. 1 (2002): 49–76; Ethan Nadelman, "Global Prohibition Regimes: The Evolution of Norms in International Society," *International Organization* 44, no. 4 (1990): 479–526.

34. See, for example, UN General Assembly, "Review of the United Nations Peacebuilding Architecture," Resolution 70/262 (New York: April 27, 2016) on sustaining peace.

35. International Crisis Group, "Libya: Achieving a Ceasefire."

14

"A Different Kind of Weapon"

Ethical Dilemmas and Nonviolent Civilian Protection

Felicity Gray

One day during my doctoral research in South Sudan, a man dies as a result of a violent altercation between civilians and armed UN peacekeepers within a Protection of Civilians site.[1] Rather than protecting the community, as they are mandated to do, it seems to many that the soldiers involved have been the source of violence and have caused direct harm. Many in the community are grieving and angry, and the death sparks many days of protest. The UN Head of Mission is chased from a *tukul*, or traditional home, where she is giving condolences to a grieving family. Other UN staff are threatened with sticks, a UN car set alight. Amid the protests, there is another death and more injured civilians. A security alert is raised, and many UN agencies and nongovernmental organizations (NGOs) suspend their activities. Food distributions and essential medical services are closed.

This incident reveals the ethical dilemma at the heart of protection of civilians: that the use of violence (even for protection) has the potential to create more violence. My research attempts to respond to this dilemma and push back against the dominance of the use of force in the direct physical protection of civilians by exploring the use of nonviolent and unarmed strategies for protecting civilians. These practices come with their own set of dilemmas: What happens if they fail? What happens when nonviolence conflicts with other principles? And finally, what dilemmas arise with attempts to integrate nonviolent practices within conventional protection frameworks? This chapter is a brief invitation to decision-makers in this space—mission architects, managers, frontline responders, and civilians themselves—to reflect on the opportunities, alternatives, and dilemmas afforded by nonviolent forms of civilian protection.

Felicity Gray, *"A Different Kind of Weapon"* In: *Wicked Problems.* Edited by: Austin Choi-Fitzpatrick,
Douglas Irvin-Erickson, and Ernesto Verdeja, Oxford University Press. © Oxford University Press 2022.
DOI: 10.1093/oso/9780197632819.003.0015

The Use of Violence for the Protection of Civilians

When a civilian population is under threat from violence, should violence be used to protect those civilians? Many argue in the affirmative. When it comes to protecting civilians from the threat of physical violence, conventional wisdom holds that the actual and potential use of violence is a necessity.[2] This use of force is justified by the idea that it is legitimate and ethical to use violence as a means of protecting civilians from harm, a key idea underpinning just war theory, international humanitarian law, and the Responsibility to Protect doctrine.

Particularly in the wake of the egregious failures to protect civilians in Rwanda and Bosnia in the early 1990s, the question of the use of force has dominated conversations around how best to protect civilians from violence. In a range of conflict settings around the world, the UN Security Council currently deploys over 100,000 peacekeepers—soldiers and military units seconded from state forces—who use assault rifles, tanks, mortar systems, and other military arsenal to respond to the physical threat to civilian life posed by a range of armed and violent actors. Though peacekeeping missions are designated as "neutral" and "impartial" and generally are authorized to use force only in self-defense (including in defense of the mandate), increasingly troops are authorized to participate in hostilities in ways that challenge their capacity to maintain these standards. As such, though humanitarian intervention aims to mitigate violence and protect civilians, it often simultaneously legitimates and perpetuates violence; in B. S. Chimni's words, it has a "dual essence."[3]

Where crises unravel to breaking points, as with genocide, immediately effective options for any kind of response—armed or otherwise—become limited.[4] I don't pretend that such situations are easily responded to with nonviolence, nor are they easily quelled by force. Speaking more broadly, in both pre-crisis and in other situations where there is undoubtedly space for alternative or complementary approaches, the use of force persists as a foundational tenet of protection of civilians, powerfully shaping thought, action, and future outcomes. There is a strong belief within protection policy circles that "a peace-related mission that is sent to an arena marked by violent conflict and that does not include a robust professional military force is doomed to failure."[5] These assumptions—often stealthily present rather than made explicit—mean that conversations around how best to respond to threats to civilian populations are narrowed.

One of the ways this narrowing occurs is by divorcing a discussion about ends—the protection of civilians and prevention of violence—from the means used to achieve them. When debate around how we should seek to protect civilians is foreclosed in this way, there is limited engagement with the potential limitations and dilemmas associated with violence as a means of protection, particularly concern that it may fuel further violence. Violent action becomes ever-present as a potential conclusion, limiting our ability to envision alternatives for action. Alex De Waal has argued that "the moment [military intervention] is canvassed, it dominates the debate like a huge dark cloud, and obscures the need and opportunity for other forms of international action."[6] In the following section, my goal is to explore what it might look like to break through that cloud and what dilemmas arise when doing so.

Using Nonviolence for the Protection of Civilians

In response to ethical dilemmas surrounding the use of violence for protection, there are organizations and individuals seeking to provide direct protection of civilians through the use of nonviolent practices. I refer to this approach broadly as "unarmed civilian protection" (UCP). Though there are many protection-related practices that could be conceived as nonviolent—such as community self-protection, for example—in this chapter UCP refers specifically to institutionalized protection practices, usually implemented or supported by an NGO, underpinned by an explicit commitment to nonviolence, in which civilians are trained to protect other civilians from violence.

UCP draws on a range of nonviolent practices—for example, protective accompaniment, protective presence, monitoring and reporting of human rights abuses, early warning and early response mechanisms—to protect civilians in violent conflict. The constellation of personnel and practices used depends on the implementing organization. For instance, the international NGO Peace Brigades International focuses specifically on third-party international volunteers providing protective accompaniment of human rights defenders. Peace Brigades International volunteers do things like accompany human rights lawyers to their offices or to court and provide presence at the offices of human rights organizations. Other organizations, such as Christian Peacemaker Teams and Witness for Peace, follow a similar model. In contrast, though the INGO Nonviolent Peaceforce also uses protective accompaniment and protective presence, they draw on a wider suite of practices in their programs, such as holding space for community safety briefings or mediation, accompanying women to collect water and firewood, and conducting shuttle diplomacy between conflicting parties with the goal of protecting civilians experiencing conflict more broadly. They employ a mix of international and national staff and see this combination as essential to their work. Other examples of UCP groups engage those local to the conflict context only, such as civilian ceasefire monitoring groups in Myanmar and Women's Peacekeeping Teams in South Sudan.

Nonviolent Ethics and Unarmed Civilian Protection

Despite some differences in implementation, commitment to a nonviolent ethic is a connecting feature among the work of UCP organizations. Though nonviolence is a principle that has multiple meanings and forms of implementation, across UCP organizations there are common ethical commitments. One example is the idea that "nonviolence goes beyond a mere refusal to carry arms. Nonviolence requires a completely different mindset that is the opposite of seeking containment, punishment and/or defeat."[7] Instead, UCP organizations argue that the ultimate goal should be to transform relationships with those who seek to harm civilians: the use of "a different kind of weapon"[8] through relationship building and using connections to both encourage good behavior and deter that which is violent or threatening. This is a stark contrast to contemporary armed peacekeeping operations, characterized by a "robust turn" toward uses of force against perceived spoilers.

This commitment to nonviolent practice relates to the underpinning ethical logic of UCP: the Gandhian notion that the means you use to achieve something cannot be divorced from the ends.[9] In other words, nonviolence is "prefigurative," seeking to align principle with practice: in popular discourse, "being the change you wish to see." Here we find an echo of the ideas advanced by Ashley Bohrer in her chapter in this volume. In practice, this means that any reliance on armed actors—for example, having an armed escort or traveling in a vehicle with an armed actor, even when you yourself are unarmed— is avoided. Instead of force, building relationships with parties to a conflict, including armed actors, is viewed as essential to the approach.[10] Practitioners are asked to demonstrate nonviolence at all times, even when off official duty.

In practice, enacting an ethic of nonviolence can be a challenge, especially in violent contexts. On some occasions, nonviolence conflicts with other key principles held by UCP organizations. What is a UCP team to do when, in their program area, it is culturally accepted and expected that a woman marry her rapist, knowing that otherwise she will be shunned by her community? In such a circumstance removing her from the violent situation conflicts with the primacy given to local actors in UCP—and potentially creates a similarly untenable and unsafe situation for the woman in question. In this case, the decision ultimately rested with the woman, who followed cultural custom despite reservations from the UCP group. In another setting, a UCP group was aware of an armed group recruiting child soldiers and faced the dilemma of whether to blow the whistle on that recruitment (knowing child recruitment would likely continue and that they would lose their access) or maintain their relationship with the armed group, knowing that this meant child soldiers remained in their ranks. Ultimately, they attempted both, maintaining their relationship and using it to place pressure on the armed group to release the child soldiers. This was ultimately effective but a much longer process of release for those children directly affected. These examples reveal nonviolent practice to be just that: a living, relational, and imperfect practice rather than a determinate principle.

Connecting with the humanity in others can be challenging when a person may have committed egregious crimes or has a gun to your head—also real-life occurrences recounted by participants in the research. This doesn't mean it is impossible or ineffective; in both these examples, all managed to use nonviolent strategies to diffuse the situation and protect themselves and others. But these kinds of security concerns do shape calculations of staff and community safety and what is considered possible in some circumstances. Tiffany Easthom, executive director of Nonviolent Peaceforce, has argued in response to regular questions about how the organization would respond to threats such as ISIS and Boko Haram, "Currently that is not work that we do, it is not something that works everywhere and in every situation. We're very careful to say that and to be humble about the reality."[11] Knowing where and when to draw this line is one of the core dilemmas facing UCP organizations. Like all actions—including armed actions—in situations of extreme violence, the consequences of miscalculation can be catastrophic. Despite these challenges, there seems to be a growing appetite among decision-makers in the civilian protection space to consider the utility of UCP practices. With this opportunity comes a new set of ethical dilemmas for UCP advocates, related to this integration. It is to these dilemmas that I now turn.

Advocating for Integration

UN mission leadership and others have begun to signal a growing openness to nonviolent, civilian-based approaches to protection. This traction in conventional protection circles thus far is evident in references to UCP in several key UN documents and fora, such as the High-Level Panel on Peace Operations report and the report of the UN Special Committee on Peacekeeping Operations.[12] In addition, the concept was included in a United Nations Security Council mandate for the first time in 2016, when it was introduced in the South Sudan mission mandate.[13] There have also been a number of protection forums at the UN focusing specifically on UCP, including an Arria formula (or informal) meeting of the UNSC and a two-day retreat for UN missions and UN Secretariat staff.

This traction comes as the result of effective advocacy by a number of INGOs who use this approach, but particularly follows efforts by Nonviolent Peaceforce to elevate this idea on the UN agenda. In the late 1990s, when founders of the organization began to advocate for investment in the approach by the UN, they recall first being ignored, then having their concerns minimized by protection policymakers who did not believe such a fringe approach had broader policy implications. Given the historical association of UCP with left-leaning civil resistance movements, this initial reluctance is perhaps unsurprising. However, in the past decade, space for advocacy of nonviolent approaches has opened as decision-makers have become more aware of limitations of armed forms of civilian protection and more open to the potential of alternative civilian-led strategies. In documents such as the High-Level Panel on Peace Operations report, the overarching narrative reflects a clear indication that armed strategies are not, and should not be, the only focus of civilian protection. Concern from member states about costs and troop commitments to peacekeeping missions and the unlikelihood of any new peacekeeping missions being approved anytime soon may also be motivating exploration of alternatives to troops-on-the-ground models of protection. Critique of the UN as an organization that has struggled to build relationships with the local communities it is seeking to serve may also be encouraging them to move toward approaches that prioritize local engagement.[14] UCP advocates have used this space to push for investment in stand-alone UCP organizations and pushed for peacekeeping troops and UN civilian staff to be trained in and utilize UCP methods themselves. This advocacy strategy reflects a belief that "recognition creates legitimacy, which creates funding,"[15] with resourcing a key factor in "scaling up" a UCP approach within influential protection institutions.

Many UN personnel observe that this pivot to taking seriously civilian capacities for protection is a work in progress, something that has been considered and implemented to disparate degrees within different mission sites. For others, mention of UCP is more likely to elicit the claim "We already do that," with officials pointing to historical examples of interposition, and civil affairs–led programs aimed at building relationships with local communities. However, often in the same conversation, it will be made very clear that this excludes activities like protection through presence or protective accompaniment by civilian staff; in the words of one UN staff member, "That is very much the purview of the military."[16] This highlights that as UCP becomes integrated into UN mechanisms and vernacular, it starts to morph conceptually. In this usage, UCP is uprooted from its grounding in nonviolent ethics and instrumentalized. It becomes a technical exercise that can be utilized within the boundaries of conventional protection rather than challenging those boundaries. The all-important connection between means and ends pursued by those practitioners with an ethical commitment to nonviolence is lost.

The Dilemma of Integration

This muddying occurs, in part, because of dilemmas faced by advocates of UCP themselves. On the one hand, advocates attempt to create distinct boundaries between nonviolent and conventional forms of civilian protection, particularly armed peacekeeping. UCP is presented as something carried out by "UCP actors" (often distinguished from armed peacekeepers, UN police, and civil affairs officers) dedicated to the particularity of the practice. At the same time, advocates call for integration of UCP into peacekeeping mission mandates and pre-deployment and in-service training for armed peacekeepers. This dual understanding of UCP reflects an underlying ethical dilemma facing advocates: there is a desire to maintain the philosophical and ethical foundations of UCP and, at the same time, to scale up the practice they so believe in. A tension arises because scaling up is contingent on resources that stem from recognition and legitimacy by institutions and states that are very focused on conventional protection logics. UCP organizations have a clear connection with nonviolence, evident in their names, their mandates, their training manuals, in interviews with staff, and across their media and communications platforms. To obtain legitimacy in those circles that can provide recognition and funding—most obviously, the UN—there can be a tendency to reframe UCP as a more technical practice, doable by anyone provided they aren't carrying a gun (rather than the approach embodying an ethical commitment to nonviolence that goes beyond the non-use of weapons). This reframing is then intensified by conventional protection actors who tend to recast UCP as a set of technical practices that are compatible and companiable with force protection, thus detaching it completely from its original ethical basis in nonviolence and civilian capacities for protection.

For many practitioners of UCP, this integration is the ultimate goal—a slow but sure spreading of ideas and practices to where they have the most potential to have an impact. Commonly, advocates point to UN civil affairs staff and unarmed UN police as actors who already implement or have the potential to pursue the approach. Another step along this continuum sees the tactics and skills of UCP as something that armed peacekeepers can and should implement. Certainly, these advocates argue, it is better that soldiers, like those involved in the civilian deaths recounted at the start of this chapter, are trained to utilize nonviolent strategies. At this point it clearly ceases to be "unarmed" civilian protection, but advocates argue that this kind of skill absorption—a soft reform of armed peacekeeping—would nevertheless improve the ability of armed peacekeepers to protect civilians. Other advocates advised the word "unarmed" be removed from the approach entirely so as to incorporate peacekeeping in an overall conception of civilian protection. Each of these moves is a step away from the original conceptual commitment of UCP to a nonviolent ethic, each step absorbing it into the civilian protection status quo. Integration thus presents a dilemma for advocates of UCP, as it risks co-optation within a status quo that is not underpinned by an ethical commitment to nonviolence in the same way.

This is a critique that has been made by UCP organizations with a more vocal commitment to solidarity (as opposed to nonpartisanship). They criticize these attempts at integration as a form of co-optation in which the ethical commitments to nonviolence and solidarity with civilians originally associated with UCP are lost. After one cross-organization UCP workshop I attended, a representative from Christian Peacemaker Teams (CPT) reflected, "The thing that I found scary was the way that neutrality was thrown around. Personally, I don't understand neutrality or nonpartisanship if you understand what racism and privilege look like on a large scale. If CPT was more neutral, we would be more well known, but I'd rather be part of a team that is proud to align ourselves with justice."[17] In essence this is a restatement of the well-known contestations around the humanitarian principles of neutrality and nonpartisanship and a testament to how these debates manifest in the decisions and work of UCP actors. As UCP grows in both recognition and diversity of approaches and priorities, this dilemma will become more and more evident.

Conclusion

How best to protect civilians in the midst of violence is a critical issue in contemporary conflicts. In the face of widespread and unmet protection needs of civilians around the world and cases of armed peacekeeping-related violence, it is perhaps unsurprising there is a growing openness to nonviolent practices like UCP. As this approach gains traction, engagement is needed around the ethical dilemmas that UCP itself faces—the way it can be in tension with other humanitarian principles and the risks facing those conducting unarmed activities. As UCP gains recognition, there will also be more and more contestation around where to draw the ethical and conceptual boundaries of the approach. If UCP is conceived too narrowly, advocates risk being unable to communicate broader applicability to civilian protection goals and actors, which limits recognition and investment in the approach and its capacity to meet protection needs. Define UCP too broadly and it loses its identity, unable to offer anything new or challenge the current boundaries of policy and practice. Attempts at palatability unravel distinctiveness and potentially undermine ethical foundations in nonviolence.

There is a need for decision-makers—both UCP advocates and leaders situated in more conventional protection institutions—to be awake to these tensions. UCP advocates need to be aware of the risks posed by attempts to fit a protection architecture that hinges on armed protection. Do that too much, and the foundation of UCP in nonviolent ethics disappears. These tensions are something that UCP field staff are keenly aware of and navigate in their day-to-day work, but that is not necessarily reflected in current advocacy strategies that aim for mainstream adoption within UN frameworks. A clear-headed discussion of the risks of particular advocacy and integration strategies needs to occur within and among UCP organizations so that these dilemmas are well considered. For those inside conventional protection institutions, particularly the UN, lessons can certainly be learned from UCP activities. The broader challenge for these actors is to step back and question some of the assumptions around the use of violence that are baked into UN protection structures to see where space might be made for deeper consideration of a nonviolent approach to protection.

UCP grows from a foundation in nonviolent ethics that is at odds with the uses of violence common to conventional direct protection practices. These ethical paradigms are not reconcilable, and so any true form of integration is likely impossible. With that said, lessons can be learned from UCP approaches that may well simultaneously better protect civilians and reduce reliance on weaponry and violence to achieve that goal. UCP reminds us that there are many ways to protect civilians—many that do not involve the use of weapons—and invites us to look for opportunities to recognize, strengthen, and invest in untapped civilian, nonviolent capacities for protection.

Notes

1. This chapter draws on ongoing doctoral research on civilian protection undertaken at the Australian National University, funded by an Australian Government Research Training Stipend and an Endeavour Scholarship. I also thank the Zolberg Institute at the New School for Social Research for their support, and John Braithwaite, Eli McCarthy, and Olivia Nantermoz for their valuable comments on drafts. In particular, this chapter draws on multisite ethnographic fieldwork and 140 open-ended interviews with practitioners and experts on civilian protection conducted between February 2018 and December 2019. Interviews were conducted in Europe, the United States, Australia, Myanmar, Lebanon, and South Sudan.
2. Though Galtung and others have highlighted that violence is a concept broader than simply the use of force (and others, such as Howard, have noted this in relation to peacekeeping), this chapter refers to the threat of direct violence in a physical sense.
3. B. S. Chimni, "Globalization, Humanitarianism and the Erosion of Refugee Protection," *Journal of Refugee Studies* 13, no. 3 (2000): 244.
4. Ken Booth, "Beyond Critical Security Studies," in *Critical Security Studies and World Politics*, ed. Ken Booth (Boulder, CO: Lynne Rienner, 2005), 275.

5. Kobi Michael and Eyal Ben-Ari, "Contemporary Peace Support Operations: The Primacy of the Military and Internal Contradictions," *Armed Forces and Society* 37, no. 4 (2011): 658.

6. Alex De Waal, "Intervention Unbound," *Index on Censorship* 6 (1994): 26.

7. Huibert Oldenhuis, *Unarmed Civilian Protection* (Geneva: United Nations Institute for Training and Research, and Nonviolent Peaceforce, 2016), 85.

8. UCP expert, teleconference interview by author, San Francisco and New York, July 17, 2019.

9. Oldenhuis, *Unarmed Civilian Protection*, 86.

10. Note there is contestation amongst UCP organizations around the degree to which relationships are built with different parties to a conflict. For example, CPT activities in Israel-Palestine take a clear solidarity position alongside Palestinians and do not actively seek relationships with Israeli authorities.

11. Tiffany Easthom, "Unarmed Approaches to Civilian Protection," statement given at International Peace Institute, New York, September 15, 2015, https://www.youtube.com/watch?v=vd3bIOFY0Fk.

12. High-Level Panel on Peace Operations, *Uniting Our Strengths for Peace* (New York: United Nations, June 2015): 37–8; United Nations General Assembly, *Report of the Special Committee on Peacekeeping Operations*, A/72/19 (New York: United Nations, 2018), 66.

13. United Nations Security Council, Resolution S/RES/2459 (2011), March 15, 2019, 2. http://unscr.com/en/resolutions/doc/2459.

14. For example, Severine Autesserre, *Peaceland: Conflict Resolution and the Everyday Politics of International Intervention* (New York: Cambridge University Press, 2014).

15. Mel Duncan, director of advocacy and outreach, teleconference interview by author, September 13, 2018.

16. UNDPO official, interview by author, New York City, September 13, 2018.

17. CPT staff member, teleconference interview by author, July 1, 2018.

15

The Ethics of Transitional Justice

Tim Murithi

As a scholar and practitioner, I have spent the past twenty-five years working in the field of peacebuilding and transitional justice in Africa. In my work, in South Africa and across the rest of the continent, I continuously come across the dilemma of how to build peace while also pursuing justice for victims and survivors who have suffered human rights violations in situations of war or authoritarian rule. The dilemma emerges because building peace requires restoring broken relationships, while pursuing retributive justice requires identifying those who are guilty of serious war crimes and subjecting them to prosecution. The dilemma arises because some of those who are identified as guilty, who tend to be leaders, are also required to proactively contribute as partners toward peacebuilding processes. The dilemma becomes a wicked problem when it becomes evident that those who have been identified as the guilty perpetrators were themselves victims of human rights violations earlier in their lives. The source of the wicked problem is the fact that, as a victim, the perpetrator should have received redress and accountability for the violations he or she suffered, and more often than not this has not happened. If the perpetrator, who was a former victim, then goes on to commit human rights violations against other people, is there a requirement for ethical action to be focused on their victimhood and to address their moral claims to justice? On the other hand, what is the ethical imperative to prosecute them for their crimes against other people? This ethical dilemma emerges repeatedly in postconflict situations across Africa, including in Burundi, Kenya, Mali, Rwanda, South Africa, Uganda, and Zimbabwe.

This chapter will discuss this trade-off between peace and justice as it relates to the African countries that are implementing peacebuilding processes. There is no "right" or "wrong" answer to the question of whether to build peace first and then pursue justice at a later stage. Each country has to approach this ethical dilemma by promoting its own internal dialogue and determining which issues it prioritizes. Even if one country chooses the

Tim Murithi, *The Ethics of Transitional Justice* In: *Wicked Problems*. Edited by: Austin Choi-Fitzpatrick, Douglas Irvin-Erickson, and Ernesto Verdeja, Oxford University Press. © Oxford University Press 2022. DOI: 10.1093/oso/9780197632819.003.0016

ethical action of building peace over the equally moral intervention of pursuit of justice (or vice versa), the wicked problem of addressing the historical
and long-standing violations of the past never disappears. Instead, it remains
a constant and unresolved tension in peacebuilding and transitional justice
processes. For example, I live in South Africa, a country that chose the path
of peacebuilding and reconciliation, but more than a quarter of a century
since the end of apartheid and colonialism, the majority of the victims and
survivors of the violations of the past have not experienced a genuine change
and transformation in their lives in terms of socioeconomic and psychosocial
well-being. Therefore, victims and survivors are now making more demands
for a more retributive and punitive form of justice against the perpetrators
and bystanders who benefited from the system of apartheid in the past.

As I noted earlier, the ethical dilemma of transitional justice is most pronounced when it attempts to treat victims as perpetrators. For example, for
the past three decades, the Lord's Resistance Army (LRA) has been in conflict with the government of Uganda. The LRA is a notorious militia that
has abducted children and forced them to join its militia forces. It is estimated that over sixty thousand children have been abducted and approx-
imately 2.8 million civilians have become Internally Displaced Persons
in camps across northern Uganda. In December 2003, the International
Criminal Court (ICC) became involved in Uganda. In June 2005, the ICC
began its case *The Prosecutor v. Joseph Kony, Vincent Otti, Okot Odhiambo
and Dominic Ongwen*. The suspects were charged with significant crimes.
For example, Ongwen was charged with war crimes and crimes against humanity, including murder, enslavement, inflicting bodily harm, looting,
and cruel treatment of civilians. The ethical dilemma emerges from the fact
that Ongwen was abducted by the LRA when he was a ten-year-old child.
Consequently, Ongwen was a child victim of the LRA militia, and therefore he is within his legal and moral rights to have justice for the violations
he suffered, including forced abduction, abuse, and trauma. However, after
many years, Ongwen eventually became an LRA commander and now led
the militia in perpetrating human rights violations, including against other
children.

The case forces us to consider several ethical dilemmas that are at the heart
of transitional justice. Specifically, is Ongwen not legally and morally entitled to some form of justice for what he endured as a child? What is the role
of justice in the larger goals of making and building peace by addressing the
socioeconomic and psychosocial harms that are endured by victims? How

do you balance the real moral concerns of victims of past violations with the need for a more humane and restorative form of transitional justice? How do you do this when so many perpetrators are themselves victims?

There are no easy ways out of these dilemmas. There are no "right" answers that universally apply. Furthermore, besides the ethical dilemma of balancing peace and justice, there are other dilemmas that emerge. Specifically, even if a retributive form of justice delivers punishment through the courts, this does not address the exploitation that people, and their descendants, suffered through slavery, colonialism, apartheid, and authoritarian rule. The effects of slavery and the loss of land through colonialism have lingering effects that are transmitted to subsequent generations. Transitional justice does not yet have a coherent response for how to deal with the historical violations suffered through slavery and colonialism. This essay suggests that the ethics of transitional justice need to move beyond addressing only human rights and political violations to also address socioeconomic and psychosocial issues, which have been endured by the victims, perpetrators, and their descendants. The African continent will continue to deal with these issues in the next decade and beyond, and therefore it will be important to continue to analyze the processes that are unfolding across the continent.

The Purpose and Function of Transitional Justice

The transitional justice approach is rooted in a legalistic tradition, with a biased emphasis on the use of judicial processes to address human rights and political violations in countries undergoing rebuilding processes. Africa's experience demonstrated that traditional notions of transitional justice needed to be rethought and reframed. In order to effectively address the real concerns of victims of past violations, African actors pushed for the expansion of the ethical parameters of transitional justice beyond their narrow human rights and political focus to include socioeconomic and psychosocial issues.

Transitional justice seeks to address past violations through processes developed to discover the truth of what happened in the past and to use this knowledge to restore broken relationships.[1] Transitional justice is usually used as a bridge to help a country restore its system of government, based on a constitution and the rule of law, in a way that manages the social, political, and economic tensions within society. Transitional justice efforts may also focus on recovering the truth of what happened through specialized bodies such as truth and reconciliation commissions and special courts. Alex Boraine argues that there are at least five components of a transitional justice process: ensuring accountability, using nonjudicial truth-recovery mechanisms, reparations, institutional reform, and reconciliation.[2] Reconciliation is understood as the cumulative outcome of the broad-based application of transitional justice processes. Concretely, reconciliation processes require that the affected parties:

1. Recognize their *interdependence* as a prerequisite for consolidating peace.
2. Engage in genuine *dialogue* about questions that have caused deep divisions in the past.
3. Embrace a *democratic attitude* to creating spaces where they can disagree.
4. Work jointly to implement processes addressing the legacies of socioeconomic *exploitation and injustices*.[3]

At the heart of reconciliation is the achievement of the ethical principles of justice and equality.[4] Consequently, transitional justice is viewed as an intermediary process that gradually and over time leads toward the promotion of reconciliation.

The Trajectory of Transitional Justice: Interrogating the Moral Contradictions

The processes that the field of transitional justice works to pursue have always existed in human societies. It was only in the 1980s that these processes were formally brought together under the umbrella of the academic term "transitional justice" after the experiences of Latin American countries as they tried to address the human rights violations of the past.[5] The Latin American approaches placed an emphasis on truth recovery as an ethical prerequisite to enabling victims and their family members to find closure for the violations they suffered. The genocides in Rwanda in 1994 and in Srebrenica in 1995 further activated the need to understand how societies that had endured mass atrocities could establish ethical processes to deal with such a brutal past and enable a society to move forward.

Similarly, in 1994 South Africa had finally been liberated from the violent system of apartheid and white supremacy, creating a new system of democracy through a constitution and the emphasis on the rule of law. The South African experience with transitional justice showed that it was also necessary to address the social, economic, and political dilemmas that the country faced as a result of apartheid. In the case of South Africa, it was a whole group of so-called White people, which is an artificial racial category, that benefited from the economic exploitation of so-called Black and Brown people. The wicked problem can be seen in the question of how white people and their descendants can collectively contribute to the restoration of the human dignity of the Black and Brown people and their descendants. Clearly, it is not possible to prosecute all the millions of white people. In fact, South Africa did not even prosecute those white leaders who were the most responsible for establishing and sustaining the apartheid system of white supremacy. The implications of these decisions resonate in contemporary South African society.

Transitional justice processes have also paid lip service to addressing gender-based violations in practice. Authoritarian rule and violent conflict affect women and men in different ways. Specifically, gender-based violence is a common atrocity that has to be considered in a transitional justice process. Around the world, women have traditionally been excluded from defining and framing transitional justice processes because political leadership around the world is dominated by men. South Africa did not focus sufficiently on addressing the apartheid violations against women. Today South Africa has an epidemic of gender-based violence and the whole society needs to proactively prevent violations against women and girls, even as it addresses the psychosocial and trauma-related issues that the rest of the population is confronting. African women and men need to ensure that the voices of women are incorporated into the design and framing of transitional justice processes. This requires the active participation of African women in the political negotiations that define the ethical orientation of transitional justice processes and the nature of its institutions.

As an overarching issue, transitional justice is understood as relating to processes that are driven by the state and state actors. Contemporary approaches to transitional justice have an almost exclusive focus on national processes. Yet the national focus of transitional justice processes is increasingly unsustainable, particularly in situations where conflicts, and their effects, spill across borders. In addition, most conflicts are increasingly cross-border in nature and are in some instances sustained by cross-border support and resources. Consequently, efforts to redress violations that emerge as a result of these conflicts are incomplete if they focus only on national actors and state-driven processes. It is necessary to find a way to pursue and promote transitional justice across borders. This raises the challenge of the prospects for institutionalizing cross-border and regional reconciliation approaches to transitional justice.

The UN Foray into Transitional Justice

The UN began to engage in transitional justice when it attempted to address the genocides in Rwanda and Bosnia. In 1997 Louis Joinet outlined the principles relating to addressing impunity in his *Final Report on the Administration of Justice and the Question of Impunity*, which he submitted to the UN sub-commission within the Commission on Human Rights.[6] Five years later, UN Secretary-General Kofi Annan issued a report entitled "The Rule of Law and Transitional Justice in Conflict and Post-Conflict Societies," which defines the field as relating to "the full range of processes and mechanisms associated with a society's attempts to come to terms with a legacy of large scale past abuses, in order to ensure accountability, serve justice and achieve reconciliation."[7] The report further suggests that transitional justice processes "may include both judicial and non-judicial mechanisms, with differing levels of international involvement (or none at all) and individual prosecutions, reparation, truth-seeking, institutional reform, vetting and dismissals or a combination thereof."[8] In 2005, the UN Commission on Human Rights revised the Joint Principles,[9] to include:

- The right to know.
- The right to justice.
- The right to reparation.
- The guarantee of nonrecurrence.

From 2005 to 2012 the UN laid out key instruments of transitional justice, including prosecutions, truth commissions, reparations, memorials, and institutional reforms.

The African Union's Transitional Justice Policy: Ethical Dimensions

Between 2010 and 2019 the African Union became the first regional organization to actively work on developing a specific policy relating to transitional justice. In February 2019, the African Union Transitional Justice Policy was formally adopted by the African Union Assembly of Heads of State and Government in Addis Ababa, Ethiopia, with the purpose of encouraging member states to broaden their understanding of justice beyond retribution to encompass restorative and transformative measures found in traditional African systems.[10] It also recommends incorporating economic and social rights and "encourages states to design reparations programs that would address the structural nature of economic and social rights violations" and that "non-state actors and beneficiaries should be encouraged to participate in such programmes."[11] The African Union has advanced an ethics of transitional justice that includes social and economic rights, thus bringing balance to the traditional focus on civic and political issues. The economic and social dimension of transitional justice processes is now emerging as a key driver of sustainable transformation for societies that have experienced violations.

The Ethics in Transitional Justice

The ethical dilemma of transitional justice, namely the tension between peace and justice, has played itself out in the majority of postconflict situations in Africa. Specifically, South Africa, Rwanda, Uganda, Kenya, Sierra Leone, and Liberia have adopted processes and institutions that have sought to address the violations of the past without allowing the potential tension that could have been generated by an orthodox approach to transitional justice to overwhelm the society and undermine efforts to build sustainable peace. South Africa's, Kenya's and Sierra Leone's truth and reconciliation commissions included amnesty provisions, which granted perpetrators a consideration of amnesty if they revealed the truth of the violations they committed. This amnesty provision tried to directly address the peace versus justice dilemma faced by these countries.

The Ethics of Sequencing Transitional
Justice Interventions

There is a perceived ethical divergence between the pursuit of peace and the requirements of justice. This means that the ethics that guide the pursuit of peacebuilding, the healing of relationships, and the restoration of human dignity are in tension with the ethics of administering justice, which are focused on the prosecution and punishment of former conflict participants. Peacemaking and peacebuilding are future-oriented in the sense that they strive to prevent violence that could recur and kill innocent civilians if conditions in the country remain unchanged. International criminal justice by definition is concerned with the prosecution of human rights violations that have already transpired through the application of due process, while upholding certain legal criteria and issuing a judgment for past transgressions. Consequently, there is an ethical tension between these two processes, and trying to undertake them in tandem can occasionally generate a moral conundrum. The notion of justice remains an essentially contested concept.[12] In fact, there are multiple ethical dimensions to justice. The ethics of practice dilemma arises in terms of which judicial action to pursue, the more formal and legalistic retributive justice or the moral and quasi-formal restorative justice interventions. The dilemma in the ethics of practice emerges from the fact that there is no moral guide as to which one of these approaches should be prioritized over the other.

The experiences in Sierra Leone and South Africa demonstrated that the debate over whether a retributive or restorative approach to justice should be deployed in the aftermath, or at the point of a conclusion of a war, has not been resolved definitively. Nor can this debate be resolved definitively because the type of justice that might be appropriate in the context of one country cannot be transplanted to another. In this regard, there is a certain degree of context-specificity in the administration of justice. A combination of retributive and restorative processes of justice can be deployed to address the needs of a society in transition.

The sequence in which either retributive or restorative justice processes are initiated is also not a precise science. In the majority of cases, retributive and restorative justice processes might be instituted and operationalized simultaneously. In some instances, the failure of a government or a society to embrace a restorative approach to justice and reconciliation can require the establishment of an international retributive/punitive justice process.

In other instances, the demands of a restorative justice process with its emphasis on truth-telling and the collective psychological transformation of promoting forgiveness and reconciliation means that efforts to administer punitive measures may need to be carefully sequenced so as not to disrupt these healing processes. In the case of the Darfur crisis in Sudan as well as in northern Uganda, individuals and leaders who have been accused of planning, financing, instigating, and executing atrocities against citizens of another group, all in the name of civil war, can be investigated by the ICC if the respective country is a state party to the ICC or if the issue is referred to the Court by the UN Security Council. In the Darfur and northern Uganda contexts, these individuals and leaders are the very same people who are called upon to engage in a peace process that will lead to the signing of an agreement and ensure its implementation.

A punitive approach to justice cannot ethically deal with the grievances that arise from slavery and colonialism and the subsequent intergenerational effects on the descendants of the victims and survivors. A punitive approach will, however, prosecute key individuals who had the greatest responsibility for committing atrocities. In spite of the available option of pursuing prosecutions for the human rights violations that were committed during apartheid, South Africa deliberately chose the moral path of placing greater emphasis on implementing a restorative transitional justice model. The option of prosecution was available during the country's transition in the mid- to late 1990s and is still viable even to this day. The fact that South Africa adopted its own unique ethical path still distinguishes it as providing an innovative model for peacebuilding. The ethical model South Africa adopted is constantly being analyzed and scrutinized for its important insights even though the country, like other countries around the world, has not yet addressed all of its challenges emanating from its past.

Sequencing in transitional justice requires establishing a coordinated approach to implementing retributive or restorative justice processes in order to ensure that stability and ultimately peace are achieved in a given country-context. For example, Tunisia and Liberia placed more emphasis on adopting and implementing a restorative model than on prosecuting perpetrators. It is necessary to adopt an ethics of sequencing, which recognizes that there is a time and a place for prosecution, and in the context of a civil war, it may not always be immediately after the cessation of hostilities between the belligerent parties. At this point in time the tension within the country tends to be uncharacteristically high, and any attempt to prosecute individuals and leaders can often, and sometimes is, seen as an attempt to deliberately continue the "war by other means" by targeting the main protagonists to a conflict. Effectively what is called for in these situations is a period of time in which the belligerents can pursue the promotion of peace. In such a situation the efforts to promote peace, including its restorative justice dimension, would have to be given ethical precedence to the administration of punitive justice. This is with a view to laying the foundations for the stability of the society. Ultimately, the perceived ethical divergence between peace and justice is based on an artificially constructed and false dichotomy between the pursuit of peace and the administration of justice.

Conclusion

This chapter has argued that the purpose and function of transitional justice must be underpinned by an ethical imperative. Several countries are emerging from conflict, and the challenge of peacebuilding is immediately confronted by the demands for justice for the victims of human rights atrocities. Traditionally, the pursuit of justice in international relations was considered detrimental to achieving peace and reconciliation, which are inherently political processes. However, Africa's experiences have challenged this presumption and demonstrated the necessary interface between transitional justice and peacebuilding processes.

Africa's experience demonstrated that traditional notions of transitional justice needed to be rethought and reframed, especially in order to address the real moral and ethical concerns of victims of past violations. The chapter argued that the dilemma arises because there is no way of determining which ethics of practice to prioritize, retributive or restorative justice. In addition, it is clear that for a qualitatively better form of ethics of practice to emerge, its norms have to be expanded beyond their narrow civil and political focus, to include socioeconomic and psychosocial issues. The essay also argued for the ethical sequencing of transitional justice interventions so that peacebuilding processes can be pursued as a moral priority to prevent further loss of life. Consequently, in line with an ethical sequencing process, the administration of justice can be pursued in a way that does not undermine the fragile peace that can further destabilize societies. The ethics of transitional justice will become increasingly relevant in a world in which an emphasis on redress and accountability for past injustices is becoming more pronounced.

Notes

1. Alexander L. Boraine, "Transitional Justice," in *Pieces of the Puzzle: Keywords on Reconciliation and Transitional Justice*, ed. Charles Villa-Vicencio and Erik Doxtader (Cape Town: Institute for Justice and Reconciliation, 2004), 67.
2. Boraine, "Transitional Justice," 67.
3. Fanie Du Toit, "A Double-Edged Sword," in *Reconciliation: A Guiding Vision for South Africa?*, ed. Ernst Conradie (Johannesburg: Ecumenical Foundation of Southern Africa, 2013), 88–91.
4. Tim Murithi, *The Ethics of Peacebuilding* (Edinburgh: Edinburgh University Press, 2009).
5. Paige Arthur, "How Transitions Reshaped Human Rights: A Conceptual History of Transitional Justice," *Human Rights Quarterly* 31 (2009): 321–367.
6. Louis Joinet, "Final Report on the Administration of Justice and the Question of Impunity" United Nations Committee on Human Rights (New York: United Nations, 1997). http://hrlibrary.umn.edu/demo/RightsofDetainees_Joinet.pdf.
7. United Nations, "The Rule of Law and Transitional Justice in Conflict and Post-Conflict Societies," Report of the Secretary-General, S/2004/616, 2004, 4. (New York: United Nations). https://digitallibrary.un.org/record/527647?ln=en.
8. United Nations, "The Rule of Law and Transitional Justice in Conflict and Post-Conflict Societies," 4.
9. United Nations, "Updated set of principles for the protection and promotion of human rights through action to combat impunity" (E/CN.4/2005/102/Add.1). (New York: United Nations, 2005).
10. African Union, *African Union Transitional Justice Policy* (Addis Ababa: African Union, 2019).
11. African Union, *African Union Transitional Justice Policy*.
12. Nick Grono and Adam O'Brien, "Justice in Conflict? The ICC and Peace Processes," in *Courting Conflict? Justice, Peace and the ICC in Africa*, ed. Nicholas Waddell and Phil Clark (London: Royal African Society, 2008), 13–20.

16

Why the Peacebuilding Field Needs Clear and Accessible Standards of Research Ethics

Elizabeth Hume and Jessica Baumgardner-Zuzik

In the peacebuilding field, there is a pressing need for rigorous research in crafting viable and effective programming. Besides the principle of conflict sensitivity, the field is not governed by an ethical code or standards outside of personal values, morals, and concepts of "doing the right thing."

This relative vacuum poses several dilemmas. As conflict experts with more than thirty-five years combined experience overseeing complex peacebuilding programming and leading diverse research efforts across the development, humanitarian, and peacebuilding nexus, we have grappled firsthand with many such challenges. As researchers, we repeatedly need to ask ourselves what trade-offs are acceptable in the pursuit of answers to our research questions *as a matter of course.* When do we rigorously follow methods at the expense of the rights of our subjects? Are we even able to im- partially determine when we are privileging our research over beneficiary well-being? How do we adjust for asymmetric power relations when working with vulnerable populations? When does the answer justify the cost—not just fiscally but, more important, to subjects' rights and moral standing? We strongly believe the peacebuilding field must develop and adopt more robust and explicit ethical codes that provide accessible standards of meas- urement and practical tools for application to help practitioners navigate these dilemmas. For too long the peacebuilding field has had a "Get out of jail free card" when it comes to design, monitoring, and evaluation (DM&E) standards. Establishing ethically informed norms and standards promotes the inherent aims of research—the search for knowledge, the utmost re- gard for truth, replicability, avoidance of error, accountability—and, most

Elizabeth Hume and Jessica Baumgardner-Zuzik, *Why the Peacebuilding Field Needs Clear and Accessible Standards of Research Ethics* In: *Wicked Problems.* Edited by: Austin Choi-Fitzpatrick, Douglas Irvin-Erickson, and Ernesto Verdeja, Oxford University Press. © Oxford University Press 2022. DOI: 10.1093/oso/9780197632819.003.0017 important, results in better programs that save lives, reduce violence, and build sustainable peace. Adhering to standardized norms will ultimately professionalize and strengthen the field, improve programming, safeguard the rights and moral standing of research subjects, and ensure organizational accountability to beneficiaries and donors.

In this chapter, we discuss the current state of ethical standards and lay out six minimum practical guidelines for ethically informed research that underscores the need for greater conflict sensitivity, higher standards for program DM&E, and stronger protections for vulnerable populations. These guidelines, we contend, can help address hard ethical dilemmas that researchers frequently confront conducting research in conflict-affected and fragile contexts.

Ethical Standards in Other Development Sectors

Ethical standards govern many disciplines and sectors that implement programs in conflict-affected and fragile states. While ethical decisions are challenging in these contexts, notable progress has been made in the humanitarian sector. For example, the Sphere Standards Humanitarian Charter and Minimum Standards (Sphere Standards) is a rights-based framework built on "the legal and ethical foundations of humanitarianism with pragmatic guidance, global good practice, and compiled evidence to support humanitarian staff wherever they work." The Sphere Handbook reflects two decades of experience from frontline operations, policy development, and advocacy to uphold principled, quality humanitarian assistance and accountability. The Sphere Project was a global movement started in 1997 by humanitarian nongovernmental organizations (NGOs) and the Red Cross and Red Crescent Movement to improve the quality of humanitarian assistance and to be accountable for their actions.[1] Significant humanitarian assistance projects in the 1990s, responding to the Rwandan genocide and subsequent Great Lakes refugee crisis and the Bosnia and Herzegovina War, were criticized by donor and NGO evaluations, and the field began exploring the idea of formulating humanitarian response standards. It was clear that "the days of unquestioning acceptance of the good work of humanitarian agencies were over. As agencies entered more difficult conflict environments, they were subject to much more rigorous scrutiny and more sophisticated political analysis."[2]

The development of standards also requires practical tools outlining their application and use. The Sphere Project provides extensive training materials focusing "mainly on the day-to-day work of the individual humanitarian practitioner" and capacity building to ensure the effectiveness and quality of programming.[3]

As the peacebuilding field matures, the development and implementation of ethical standards must match the field's specific goals in a practical and meaningful way, especially to facilitate the adoption of best practices and safeguard rights and protections. These standards need to encompass the principles of conflict sensitivity and human subjects research, while also supporting impact, embracing failure, and protecting the dignity, rights, and well-being of its beneficiaries. The need could not be more significant, especially as the number of highly violent conflicts increased in 2019 for the first time in four years,[4] and roughly 2 billion people live in countries where development is affected by fragility, conflict, and violence.[5] The COVID-19 global health pandemic is "stabilization in reverse," further accelerating conflict dynamics in fragile states.[6]

The State of Peacebuilding Research Ethics

Searching for resources on peacebuilding ethics returns few results because there are currently no established industry standards, although there are a few excellent guides available on reflection and theory. In *Ethics for Peacebuilders: A Practical Guide*, Reina C. Neufeldt focuses on peacebuilding ethics and morality and begins to tie peacebuilding ethics to practice, arguing for an adaptive management approach in the program implementation cycle.[7] Adaptive management requires more flexible systems—both programming and DM&E—that allow for experimentation and adaptation of program activities, plans, and strategies. Neufeldt discusses a dynamic ethics reflection-action cycle, which recommends one assess, inquire, deliberate, decide, watch, and adjust, but stops short of outlining practical standards. She further makes a case for developing a community of practice through the Alliance for Peacebuilding (AfP). In her reflection piece "10 Ethics of Peacebuilding," Lisa Schirch suggests putting the dignity and humanity of all people at the center of all that one does, talking and meeting with all parties to a conflict, and being ready to change oneself.[8] While these resources provide guidance and reflection on ethical values, morals, and dilemmas, they do not address ethical concerns related to peace research.

Minimum Standards for Peacebuilding Research Ethics

Ethical standards for DM&E and research with human subjects have a long history deeply rooted across multiple sectors, particularly the health field.[9] Research ethics are critical for all sectors collecting data and conducting research, including the peacebuilding field. We argue that while peacebuilding programming may not align with clinical research, basic standards of research ethics must be integrated within the field and are inherent in its very ethos. There are many examples where the rights of subjects have been manipulated and flagrantly disregarded in the pursuit of research—a threat even more relevant to a field that is already operating in a context of upheaval, threatened livelihoods, and political violence.[10]

To address this gap, we recommend the adoption of minimum standards: applying conflict sensitivity to all aspects of programming, establishing robust DM&E, securing informed consent, adhering to best practices for ethics approval of programming, upholding data protection and privacy rights, and transparently sharing results—both successes and failures. We discuss these in more detail next.

Conflict Sensitivity

In 1993, CDA Collaborative Learning launched the Local Capacities for Peace Project, which found that in conflict settings, "aid can reinforce, exacerbate, and prolong the conflict; it can also help to reduce tensions and strengthen people's capacities to disengage from fighting and find peaceful options for solving problems."[11] AfP treats conflict sensitivity as an overarching standard that applies to all aspects of DM&E. Conflict sensitivity is a set of principles for operating in conflict environments that requires organizations to (1) understand the conflict dynamics in the areas in which they work, (2) understand how their interventions interact with those dynamics, and (3) take steps to ensure that their actions reduce negative outcomes and increase positive outcomes.[12] However, even though conflict sensitivity has been around for more than two decades and is widely accepted by all major government donors and NGOs, many organizations and practitioners struggle to integrate it into their programming.

Design, Monitoring, and Evaluation

Too much is at stake in fragile and conflict-affected settings to not understand the impact of peacebuilding programming, and yet many organizations do not rigorously evaluate the impact that their programs are having on the ground. The field needs to better understand and improve its impact, which requires a field-wide cultural shift that values transparency, embraces open inquiry, and shares knowledge from successes and failures. Peacebuilding practice will be improved by ethically informed DM&E that designs evidence-based programming, monitors programs for ethical abuses and accountability, and evaluates the impact of programs on beneficiaries.

The peacebuilding field has made strides in addressing the technical challenges of peacebuilding evaluation by developing innovative strategies to measure and learn. Incremental improvements in methodological rigor include enhanced opportunities for shared learning and fostering the use of evidence to inform policy and practice. However, much of this progress has been made by individual organizations, not at the field-wide level, and donors have yet to prioritize and fund it effectively. The field continues to struggle with DM&E and faces substantial obstacles in answering the following questions: What brings about lasting and sustainable peace? For whom? and In what context?

AfP works to end conflict and build sustainable peace worldwide and is leading the peace and conflict field to embrace a more adaptive and rigorous evaluative culture to prove impact. The field demands leadership to unify a dynamic yet disparate field to break down silos within the community; build bridges to policymakers, other sectors, and the private sector; and raise the bar of peacebuilding practice toward shared learning, adoption of evidence-based approaches, and greater impact.

Through our Learning & Evaluation program, we focus on building capacity to capture high-quality, actionable data, encouraging the field to become more evidence-based, and making a stronger case for peace through shared research efforts to synthesize impact. Through these efforts, we believe peacebuilders will be more effective collectively.

In 2011, AfP began leading the Peacebuilding Evaluation Consortium in partnership with CDA Collaborative Learning Projects, Mercy Corps, and Search for Common Ground, a field-wide effort to measure and learn from peacebuilding programs. The Consortium started by asking *What are the barriers for better DM&E in the peacebuilding field?* and *Why is there such hesitation and resistance to adopting DM&E basic standards?*

The Consortium developed practical resources that contribute to strengthened approaches to DM&E. In 2016, it conducted a closed-door roundtable discussion with representatives from U.S. government agencies, peacebuilding practitioners, and evaluation experts to determine if it was possible to develop and define standards for DM&E. Participants were split on the need to develop standards. The Consortium has ended, but in 2018 we published a document titled "Guiding Steps for Peacebuilding Design, Monitoring, and Evaluation."[13] This document details seven steps that are a minimum to contribute to robust evidence and learning in the peacebuilding field. Each step outlines critical elements and provides an initial list of key resources supporting them: (1) conduct a conflict assessment; (2) design a peacebuilding program; (3) develop a monitoring and evaluation plan; (4) conduct a baseline study; (5) develop monitoring and adaptive management; (6) conduct an end-line study and/or final evaluation; and (7) disseminate and share results and key learnings. The seven steps cover *what* an organization should do, but not *how* it should be done.

However, we believe it is time to develop standards for DM&E and peacebuilding research, moving beyond the "Guiding Steps" and holding the field accountable to ethical research practices. These minimum standards must incorporate practical guidance, establish global best practices, and provide evidence to support peacebuilding organizations and staff in assessing when the risk outweighs the benefits of an intervention. Central to these standards must be the protection of vulnerable populations through the adoption of recognized research best practices, including informed consent, institutional review boards and ethics committees, and data protection and privacy rights.

Informed Consent

Informed consent is an ongoing process whereby people are informed about the proposed programming and research and which allows them to make a voluntary decision about whether they wish to participate. It protects one of the fundamental ethical principles of autonomy: where responsibility to participate must be given to the individual. Obtaining informed consent involves multiple steps to inform potential participants about the purpose of the program, the specific methodology, potential risks and benefits of participating, and participant rights.

Three fundamental aspects of informed consent include voluntariness, comprehension, and disclosure.[14] Individuals must willingly decide to participate and should not be influenced by anyone involved in programming or research. Individuals must have the mental and decisional capacity to understand the information presented. Finally, all potential risks and benefits of participation must be disclosed to potential participants.[15] These principles are particularly relevant when working with vulnerable populations, including prisoners, ex-combatants, refugees, internally displaced peoples, migrants, children, and pregnant women, making the concept of informed consent not an option but a necessity for peacebuilding programming.

The informed consent process is described in many ethical codes and regulations for human subjects' research, including the U.S. Health and Human Services Office for Human Research Protections[16] and the European Commission's Research Directorate General.[17] However, it is commonly required only when conducting research in collaboration with academic institutions or specific donors. We argue that any time an organization collects data—whether to assess program effectiveness or donor accountability or to evaluate programmatic hypotheses—research is being conducted. Therefore, researchers must adhere to the principles of research ethics. Integrating best practices of informed consent improves the quality and integrity of research and guarantees that the dignity and human rights of participants are at the forefront of research considerations.

Institutional Review Boards and Ethics Committees

Institutional Review Boards (IRBs) and Ethics Committees (ECs) are administrative bodies established to protect the rights and welfare of human subjects for medical or other research studies. They carry out ethics reviews of proposed research involving human participants to approve, disapprove, monitor, and modify these efforts to guarantee the safety and well-being of participants. IRBs and ECs are often located within academic/research institutions or government bodies at the local, national, or regional levels. They were developed in direct response to research abuses occurring throughout the twentieth century and stem from the publication of the Belmont Report[18] and the Nuremberg Code,[19] which outline sets of research ethics for human experimentation and protection of participants in clinical trials and research studies.

All research involving human subjects should be reviewed by an EC to ensure appropriate ethical standards are upheld. The U.S. Health and Human Services Office, the European Commission, and most accredited academic institutions require IRB and EC approval for all research proposals submitted, but grantmaking and funding organizations outside of these governmental and academic bodies do not require the same.

The peacebuilding field is lagging behind more professionalized fields in developing an evidence base, which is partially due to the lack of IRBs and ECs for peacebuilding programming and research. Some progress is being made through heightened collaboration with academic institutions and the development of internal IRBs—including at the U.S. Institute of Peace. Increased cross-sectoral and academic collaboration will provide greater opportunities to employ IRBs and ECs within the peacebuilding field; however, leading funding institutions must require greater ethical oversight for peacebuilding programming and research. Additionally, peacebuilding organizations ought to develop more robust internal/interdepartmental IRBs and ECs adhering to industry standards that are accountable to national and international ECs in the regions and countries where they work. While ECs and IRBs can pose additional dilemmas, such as an often top-down approach to research, institutionalized standards of ethics, and treatment of research participants as subjects rather than collaborators, we contend that the risk of not having them poses a much greater threat than these individual challenges. Developing minimal standards will allow the peacebuilding field to contextualize and adapt these protocols from other sectors to the field's specific needs and address some of these inherent dilemmas. Potential adaptations could include more inclusive ethics review boards that involve researchers, implementers, and regionally representative participants.

Data Protection and Privacy Rights

The landscape of research and data collection has significantly transformed in the past two decades, with the advent of new and innovative tools allowing the field to collect, access, and use data in ways previously unimaginable. With these opportunities come many new risks and responsibilities to safeguard beneficiaries' data and privacy and avoid dual use.

Privacy and data protection are fundamental rights of all people,[20] but progress is needed to guarantee these rights, particularly in development and peacebuilding programs occurring in international contexts. Critical considerations for peacebuilding research ethics relate to identifying risks to privacy, balancing known and unknown risks with the benefits of open and transparent data, and avoiding dual use of research in fragile environments. Protection of peacebuilding data is as sensitive as health data because it includes highly sensitive information related to violent extremism, trauma, sexual and gender-based violence, and human rights abuses.

Identifying risks and guaranteeing beneficiary privacy is critical to upholding conflict sensitivity in peacebuilding.[21] A critical first step is requiring informed consent. Additionally, best practices of storing and protecting personal and geographically identifiable information (PII) must be instituted across the field. These practices include the use of unique identifiers for all research projects, separation of PII from data for hard-copy storage, requiring human subjects' certification for anyone handling PII, and the use of encryption protocols for storing and transferring data.

While currently many of these best practices are only recommendations, the advent of the European Union's General Data Protection Regulation (GDRP) is quickly turning them into law.[22] The GDRP developed rigid groundwork for the protection of personal data that has influenced major changes in data protection regulation globally.

Peacebuilding programming must address key considerations for balancing known and unknown risks and benefits of data collection and use. There is a delicate balancing act when identifying the data an organization needs to collect versus the data it wants to collect. This consideration should be based on organizational and programmatic needs and the benefits of the data compared to the risks associated with collecting data. The U.S. Agency for International Development recently released the "Considerations for Using Data Responsibly,"[23] identifying the risks associated with development data, which is relevant to peacebuilding. Conducting a risk-benefit analysis across programming data has great potential for improving data protection and ensuring privacy rights.

A final ethical dilemma for peacebuilding research relates to dual-use research of concern, which is research that can be reasonably anticipated to generate knowledge or information that could be directly misapplied to pose a significant threat.[24] The threat of dual-use research of concern is particularly relevant to the peacebuilding field due to the sensitive nature of its data and its potential for misuse by parties to the conflict and those collecting it in fragile contexts. Special measures must be taken to protect data collected to ensure the data do not support one of the conflicting parties over the other and to prevent misappropriation and use by any parties.

Conclusion

The development of peacebuilding research ethics is critical to governing the standards of conduct for research in conflict-affected and fragile states. The peacebuilding field requires more than good values, morals, and principles to ensure their programs and policies do not exacerbate conflict dynamics and guarantee the safety and well-being of participants. The research standards presented are the minimum that need to be adopted. As such, they represent only the beginning of a larger process.

The peacebuilding field, like all professional sectors, requires more than "doing the right thing" to govern practice. We should not be exempt from following industry standards and best practices for research ethics to ensure compliance, accountability, and responsibility for our work. If we believe what we do is important, then we need to ethically prove it. Peacebuilding research ethics are inherently fundamental to the ethos of the peacebuilding field, as exemplified by other sectors working in similar contexts. Additionally, there should be accountability for noncompliance by donors, academics, implementing organizations, peers, and program beneficiaries.

If the intent is just, it does not necessitate the means will inherently be so—whether intended or unintended. Establishing and implementing peacebuilding research ethics to govern the standards of conduct for research is not only a moral imperative but a best practice to safeguard the dignity, rights, and well-being of the populations we serve.

Notes

1. The Sphere. https://spherestandards.org.
2. Margie Buchanan-Smith, "How the Sphere Project Came into Being: A Case Study of Policy-Making in the Humanitarian Aid Sector and the Relative Influence of Research," Overseas Development Institute, July 1, 2003, https://cdn.odi.org/media/documents/176.pdf.
3. The Sphere. https://spherestandards.org/about.

4. United Nations Office for the Coordination of Humanitarian Affairs, "Global Humanitarian Overview 2020," December 10, 2019, https://www.unocha.org/sites/unocha/files/GHO-2020_v9.1.pdf.

5. Organisation for Economic Co-operation and Development, *States of Fragility 2018* (Paris: OECD Publishing, 2019, https://doi.org/10.1787/9789264302075-en. See esp. chapter 3, "How Are Fragile Contexts Faring in Achieving Sustainable Development?"

6. Liz Hume and Conor Seyle, "COVID-19 Is Stabilization in Reverse: Applying the Principles of the Global Fragility Act to Pandemic Response," Alliance for Peacebuilding, May 8, 2020, https://medium.com/@AfPeacebuilding/covid-19-is-stabilization-in-reverse-applying-the-principles-of-the-global-fragility-act-to-6bd3baba4878.

7. Reina C. Neufeldt, *Ethics for Peacebuilders: A Practical Guide*, (Lanham, MD: Rowman & Littlefield, 2016).

8. Lisa Schirch, "10 Ethics of Peacebuilding," *Lisa Schirch* (blog), May 17, 2018, https://lisaschirch.wordpress.com/2018/05/17/10-ethics-of-peacebuilding/.

9. Marcello Ienca, Agata Ferretti, Samia Hurst, Milo Puhan, Christian Lovis, and Effy Vayena, "Considerations for Ethics Review of Big Data Health Research: A Scoping Review," *PLOS One* 13, no. 10 (October 2018): 15, https://doi.org/10.1371/journal.pone.0204937.

10. For additional review and reference, please refer to Nazi criminal and research atrocities, the 1932–1979 U.S. Tuskegee Syphilis Study, the 1963–1966 Willowbrook State School Hepatitis Experiments, the Milgram Experiment, and the 1971 Sandford Prison Experiment.

11. Mary B. Anderson, *Do No Harm: How Aid Can Support Peace—or War* (Boulder, CO: Lynne Rienner, 1999).

12. CDA Collaborative Learning Projects, "Do No Harm Workshop Participants Manual," December 2016, https://www.cdacollaborative.org/publication/no-harm-workshop-participants-manual-2016/.

13. Peacebuilding Evaluation Consortium, "Guiding Steps for Peacebuilding Design, Monitoring, and Evaluation," 2018, http://www.dmeforpeace.org/wp-content/uploads/2018/06/PEC_Guiding-Steps-for-Peacebuilding-DME.pdf.

14. U.S. Department of Health and Human Services, Office for Human Research Protections, "Guidance," February 16, 2016, https://www.hhs.gov/ohrp/regulations-and-policy/guidance/index.html.

15. U.S. Department of Health and Human Services, Office for Human Research Protections, "Guidance."

16. U.S. Department of Health and Human Services, Office for Human Research Protections, "Guidance"; Protecting Human Research Participants, "Protecting Human Research Participants: PHRP Training," 2018. https://phrptraining.com/.

17. European Commission, *Ethics for Researchers* (Luxembourg: European Commission, 2013).

18. U.S. Department of Health and Human Services, National Commission for the Protection of Human Subjects of Biomedical and Behavioral Research, "The Belmont Report," April 18, 1979, https://www.hhs.gov/ohrp/sites/default/files/the-belmont-report-508c_FINAL.pdf.

19. U.S. Department of Health and Human Services, National Institutes of Health, "The Nuremberg Code," https://history.nih.gov/display/history/Nuremberg+Code.
20. European Union, "Consolidated Versions of the Treaty on European Union and the Treaty on the Functioning of the European Union," Publications Office of the European Union, 2016, https://www.consilium.europa.eu/media/29726/qc0116985 enn.pdf.
21. Anderson, *Do No Harm*.
22. European Parliament, and Council of the European Union, General Data Protection Regulation, Law 119, 88, 2018, https://ec.europa.eu/info/law/law-topic/data-protec tion_en.
23. U.S. Agency for International Development, "Considerations for Using Data Responsibly at USAID," May 1, 2019, https://www.usaid.gov/responsibledata.
24. National Institutes of Health, Office of Intramural Research, "Dual-Use Research," September 12, 2020, https://oir.nih.gov/sourcebook/ethical-conduct/special-resea rch-considerations/dual-use-research.

17

Consent, Inclusivity, and Local Voices

Ethical Dilemmas of Teaching Peace in Conflict Zones

Agnieszka Paczyńska and Susan F. Hirsch

Between us, we've been teaching in the peace and conflict field for over three decades. Although our research and practice are in distinct subfields, together we've developed a bit of a side gig in experiential learning. While creating scholarship and materials for teaching, we've always reflected on the many ethical dilemmas that confront teachers in our field.[1] Recently, our reflection on ethics has deepened, as we've noticed a shift in our students. Increasingly, they come to the study of peace and conflict as activists, striving for particular forms of social change and predetermined approaches to achieving it. As instructors, we face a challenge: How should we teach ethics to budding practitioners who want to change the world and who have a strong sense of how it should be done?

Teaching in the Era of Black-and-White Ethics

Tried and true instructional methods such as teaching theories of ethics and ethical practice, along with acquainting students with case studies of ethical dilemmas, remain great starting points. Yet when students seek to become professional practitioners, they benefit from more engaged and intentional encounters with ethical dilemmas. As we've argued elsewhere, experiential learning both in supervised classrooms and in field-based courses provides students with closer-to-actual experiences of intervening in conflict settings and thereby confronts them with various ethical dilemmas that come up in practice.[2] As part of a broader project on experiential learning in the peace and conflict field, we developed and tested models for field-based courses in conflict zones in which students learn to become practitioners.[3] Our approach focuses on creating sustainable partner relationships with

Agnieszka Paczyńska and Susan F. Hirsch, *Consent, Inclusivity, and Local Voices* In: *Wicked Problems*. Edited by: Austin Choi-Fitzpatrick, Douglas Irvin-Erickson, and Ernesto Verdeja, Oxford University Press.

nongovernmental organizations (NGOs), universities, or other institutions and then training our students to deliver conflict resolution programming in consultation with the partners and under faculty supervision. These courses surface multiple ethical concerns.[4] One major ethical issue is that students come to the peace and conflict field because they want to do good in the world, yet, as Reina Neufeldt argues in this volume, such moral clarity can lead practitioners to not question whether the choices they make are ethical. Another major ethical issue is that students are inexperienced, and this can compromise the successful delivery of what we've promised our partners and what they expect. At the same time, such courses are designed for the practice and improvement of students' abilities and should strike the right balance of activities so that students are pushed beyond what they've previously done. In making sure that the experience engages students in learning and practicing something new, as instructors we have had to be ready to step in to ensure the effectiveness of any activity and to make certain that the whole venture doesn't go off the ethical rails. But this of course exposes the power differentials between us and the students, revealing the tensions as we try to work collaboratively with them yet remain in charge.

These courses sit at the intersection of teaching and practice, given the commitments that we as instructors have to partners and also to our students. Thus, the many ethical dilemmas that emerge are rooted in a particular educational space that, ideally, should conform to both the ethics of teaching and the ethics of conflict resolution practice. Yet being in and acting in this space sometimes makes our students uncomfortable. As our examples will show, ethical uncertainty—ethical gray areas, if you will—can emerge. These spaces of uncertainty can be especially troubling to students when their strongly held commitments are challenged.

In the era of #MeToo, Black Lives Matter, and Extinction Rebellion, many of our students hold crystal-clear ethical commitments that chart their political and social justice aims and approaches. Choices around certain issues are presented as starkly black and white. For example, many campuses are debating how to respond to students' requests that they be protected from encountering offensive or otherwise "triggering" material in their coursework. Some of the requests for warnings or avoiding retraumatizing content leave little room for discussing certain topics in class, such as sexual violence or racial discrimination. In response, Brian Leiter has made a controversial though thought-provoking assertion that trigger warnings make sense in only a narrow set of circumstances, for instance in cases of documented posttraumatic stress disorder. Otherwise, instructors should not shield students from engaging with all aspects of human experience, much of which involves gray areas, and subjects, such as oppression and marginalization, that are upsetting and likely to make students uncomfortable, albeit some more than others. As Leiter argues, it is by exposure to all aspects of the human condition that students learn and grow.[5] We share Leiter's view that just because a learning experience is discomfiting to some does not mean that students should be shielded from it. Yet such situations do require instructors' careful consideration, especially when the triggers affect students who have been marginalized or traumatized in educational settings or elsewhere.

A stark view of ethical choices is appropriate for some circumstances. For instance, the ethical commitment to "do no harm" is a bedrock of the peace and conflict studies field.[6] However, in many instances of practice, ethical gray areas are more likely than black-and-white options. Our aim, frankly, is to engage students in ethically challenging activities so that they will be equipped to recognize these gray areas and be comfortable enough to inhabit them for the time it takes to develop strategies to operate within them. To accomplish that we need to be sure that black-and-white ethical choices are not their only options, especially on issues that matter or in situations that they might later confront as practitioners. Through our teaching we encourage students to (1) put their ethical commitments into historical and cultural perspective, (2) expand the ethical approaches with which they are familiar, and (3) practice flexibility—personal and professional—in handling ethical dilemmas. Each of these strategies works against students adopting a rigid mindset that circumscribes options. Yet as instructors we need to be mindful of our own ethics of teaching, as we have to consider whether it is ethical, on our part, to attempt to disrupt, change, or directly challenge our students' ethical commitments. We know from our research that doing so through experiential learning deepens the lessons learned while also deepening the emotional response, including feelings of discomfort. This predicament is especially tricky as our positioning as instructors gives us the upper hand in relationships of power framed by required assignments, grades, and institutional authority.

As shown in examples from our teaching, the clash between students' preference for ethical clarity on the topics of consent, inclusivity, and local voices and our pedagogical strategy of immersing students in ethical gray areas sits at the heart of the new ethical dilemmas that we've been experiencing as instructors. It's not always easy to know how much to focus on the sensitive issue of ethics, how best to accompany students into the ethical gray zones that may cause them discomfort, and whether we should push back against resistance, especially in areas where students' commitments are strongly held. In most instances we found that the attempt to teach students to navigate the gray areas of practice led to ethical dilemmas for us that were not confined to the relationships between students and teachers alone but emerged in relationships to partners and thus challenged us as practitioners and instructors.[7]

Discomfort in the Gray Areas of Pedagogy and Practice

Our examples are drawn from experiential learning exercises that we and our colleagues delivered to graduate and undergraduate students in field-based contexts in the United States, Colombia, Liberia, and Malta. In these courses our students develop, implement, and evaluate programming, which has included organizing and running a dialogue, or World Café, conducting a conflict assessment or focus group, and facilitating a meeting or a conflict skills workshop, among other activities. In determining how to meet partner requests through field-based course activities, we've had to be mindful of what students are able to accomplish. In our experience, delivering programming through a field-based course necessarily involves compromises and trade-offs, as well as significant effort beforehand and reflection afterward on the part of instructors. At the same time, we are convinced that these courses offer unparalleled opportunities for deep learning.

Consent

As Elizabeth Hume and Jessica Baumgardner-Zuzik make clear in their chapter in this volume, the issue of consent is fundamental to the ethical conduct of peacebuilding work. Students can have a strong sense of what it means to consent to engage in any activity. The right to consent to participate in classroom activities, or not, is increasingly asserted by students. In deciding whether to consent, their comfort is of concern, and in working with local partners students believe that partners should have the same ability to avoid discomfort. This concern with not placing themselves and partners in

situations where discomfort might result clearly emerges when students organize Dialogue and Difference events, which are facilitated discussions on
current, controversial, or divisive topics aimed at engaging participants in
listening to views that may fundamentally differ from their own. By design,
discussions place participants in a context where there is a high probability
they'll experience discomfort. Adding to the challenge is that the students
organizing the event don't know until it begins who will participate. When
topics tap into difficult, even traumatic experiences, the issues of consent and
students' desire to avoid discomfort are heightened. This was the case when
students organized a Dialogue and Difference event titled "The #MeToo
Movement: Gone Too Far? Not Far Enough?" Following presentations by
three speakers, as participants readied themselves to engage in small-group
facilitated discussions, one of the attendees stormed out. Agnieszka spoke to
the woman, who was shaken and upset. Mentioning that she was a victim of
sexual assault herself, the attendee said that she felt erased and silenced by
the speakers, none of whom spoke to her experience as a white, middle-class
woman. During the activity debrief, the students organizing the event raised
the ethical dilemma of whether they had made the right choices in managing
such a painful discussion and whether, in their goal of ensuring a diversity of
voices, they had inadvertently excluded others, a point we address in the next
section. They also worried about how they should respond in the moment
when participants express distress and whether they could have prevented
the discomfort that led to the attendee revoking consent to participate. The
discussion included the helpful observation that perhaps by deciding to at-
tend the event, participants are giving affirmative consent to be moved in
some way by the content.

Consent is always sought before running a focus group, which is one of
the activities that we engage in with local partners in field-based courses.
Students plan and run the focus group and then analyze the data collected as
part of an assessment or research project. To prepare a class of students to run
focus groups, Susan led a model focus group and then a debrief that included
a discussion of how they would mount a focus group with the local partner,
which in this instance was a large class of law students at the University
of Malta. During the preparation students asked Susan if the law students
would be asked to give their consent to participate. Although Susan assured
them that they would, some students expressed concern that the law students
would be unable to consent if they did not know the exact questions that
would be asked in the focus group. "What if something we were planning to

ask is a trigger for one of the law students?" "Shouldn't we share the questions with them beforehand so that they can decide?" In discussing this option, Susan reviewed the research purpose of a focus group and explained that disclosing the questions could result in very different data. The discussion touched on issues that are key to ethical practice, and some students asserted that they would never feel comfortable "imposing" questions on anyone who could not fully consent beforehand. Another student ventured that consent might be understood differently by students from different cultural and political contexts and that it was inappropriate to assume too much about what might concern the law students. Comments pointed to gray areas in the ethics of consent, and students were encouraged to use the opportunity of the focus group to explore their own understandings of consent and how their assumptions about it might have been challenged.

Inclusivity

At the request of our local partner, an NGO focused on serving the migrant community, we agreed that our students would facilitate a small group discussion of mental health challenges for migrants as part of a larger event celebrating migrant women. Over the past year the attendees had been to various trainings around mental health awareness, and this activity was designed to assess their understanding of the mental health challenges in their communities and to identify directions for additional training. It was anticipated that the majority of attendees would be women and that some sensitive issues (e.g., sexual harassment or abuse) might arise during the discussion. Susan prepared the students to handle the sensitive areas and assigned an instructor or NGO leader to each table to step in if needed. Students were tasked with leading the table discussions and taking notes. We faced a challenge in that six of the twenty students were male, and, in conversation with the NGO, we determined that the presence of men in the table discussions might discourage women participants from speaking up. The students debated what should be done and decided that the male students would sit together at one table and have their own discussion. So that the male students would have an opportunity to facilitate, they were tasked with conducting a brief evaluation of the event following the table discussions.

Although they understood why, many students were upset that the men were segregated. They felt it was unfair that a key aspect of the course could not be shared equally by everyone. The women students felt guilty about their positive experience conducting the discussions, even though the male students had what they termed to be an extraordinary conversation among themselves about mental health, masculinity, aggression, and other topics. Afterward the students realized that they might not have been able to achieve the partners' goals for the event without gender segregation, yet they still felt regret. In reflecting with them, Susan drew a parallel between their uncomfortable experience and situations where practitioners determine who on a project team can best accomplish the project aims. Some participate and some stand down, and uncomfortable feelings might result.

The underlying concern that everyone in class should have the same experience returned during another field-based course in Malta a couple of years later. This time the students were tasked with running World Café activities with participants from a migrant-led NGO. This diaspora group, a longstanding partner, was trying to determine whether and how to have influence in the political shifts in their country of origin. The group leaders were familiar with World Café and encouraged the students to develop a creative approach. As they prepared the World Café prompts, the students raised a concern that participants might not want to speak in English. Group leaders assured us that most people would speak English but that they would translate if necessary. At the World Café, the questions posed by the students led to intense and constructive discussions, and sometimes participants spoke in Arabic. Designated translators were swept up in the discussion, and for periods of time the students were left out. Afterward students expressed frustration. Susan explained that they should realize that their facilitation efforts had accomplished the goal of stimulating intense discussion among all the participants, as requested by the local partner, and that their own experience of being excluded, though uncomfortable for them, had little bearing on the activity's overall success. Again, learning how to stand aside in a practice situation might be useful in professional work.

Local Voices

Students are eager to engage with local people and contribute to empowering their voices. Yet increasingly conflict resolution and peacebuilding practitioners have been questioning their own assumptions about who is perceived to be an authentic local interlocutor and whether some local voices are privileged while others, often those of less powerful or marginal-ized community members, are silenced. One challenge students encounter is that there is no sole local voice but rather a plethora of voices that reflect a community's complexity. As instructors we have become acutely aware of the ethical dilemmas around local voices and, as the next examples illus-trate, have worked to ensure that our courses are designed in such a way that the local does not become fetishized, tokenized, or exploited. An essen-tial component of that design is the reliance on long-standing relationships with local partners.

Appreciating local complexity can be especially difficult for students when working in unfamiliar environments and under time pressures, both of which generate high levels of stress. In our experience, when students experience stress, they tend to simplify, and this can generate wicked problems.[8] Students routinely make assumptions about the relationships among people they encounter and the conflict dynamics they seek to un-derstand as quickly as possible, which in turn makes it easier for them to believe that they comprehend the local situation. At the same time, their desire to empower local partners can mean that they are inclined to be-lieve those whose views they hear first. Both tendencies can mean that they come to conclusions about a conflict dynamic without ensuring that mul-tiple voices and views are heard. This dynamic was evident when students on a field-based course in Liberia were asked by our partner, a Liberian NGO, to conduct a conflict assessment in a community a couple of hours outside of Monrovia. The county land commissioner asked the NGO for the assessment, as the land conflict had been simmering for some time and had divided the town's population. The first interview the students conducted was with the land commissioner himself, who explained the conflict. When Agnieszka and the students left his office, the students felt confident they understood the conflict dynamics. But did they? Or did they now have a set of assumptions about what was driving the conflict and who was in conflict with whom? The students at first seemed confused by this distinction between assumptions and facts. After interviews with multiple residents, a more complex picture of the conflict emerged, one that had little resemblance to the land commissioner's depiction. For students it be-came clear that by jumping to conclusions they had not appreciated the conflict's complexity. They recognized that the views of one observer (or was he really a party to the conflict?) were just that: one view based on that person's positionality and interests. They also recognized that their early agreement with their local partner's client was a way to show respect toward their partner. Yet if they had followed their initial instincts, the conflict assessment they would have produced would have been inaccurate, generated faulty intervention recommendations, and potentially made the conflict worse, thus violating the first rule of conflict resolution work: do no harm. In other words, not listening to multiple voices, including those of community members who may find it challenging to speak up in open community fora, for instance women and youth, had ethical implications. This experience proved uncomfortable for the students, as they had to confront how their desire to empower local voices clashed with their urge to quickly understand the community conflict and how this clash could have resulted in ethically problematic privileging of the voices of more powerful townspeople while excluding other local voices.

Another strategy that we routinely use to help our students appreciate the complexity of local voices is to connect them with their peers. Working together with students from universities in our field-based sites provides invaluable opportunities for our students to undertake conflict resolution practice with students who represent local voices and also, potentially, different understandings of the ethical dilemmas of practice. We saw this during the field-based course we held in partnership with Javeriana University in Bogotá, Colombia. Students from the two universities lived together in tight quarters and grappled with tensions that arose from the behavior of the American students in public spaces. At the time, insecurity in the area where the students lived was high, with various armed factions entering the city at night. The Colombian students felt that their American counterparts underappreciated the security concerns and that, by making politically charged remarks in public, they could put the whole group in peril. Together the students figured out how to hear one another on these important issues and how to navigate their differences and develop working relationships based on trust. They did so by collaboratively organizing a day-long festival on alternatives to violence for a local community plagued by displacement and violence. As peers, students' ability to challenge one another on ethical matters—for instance, to exchange views about how to operate in situations of violence and security—is free of the power dynamics that we as instructors interject and thus invaluable learning in an ethical gray area.

Conclusion

As we seek to teach students to handle ethical dilemmas, we cannot promise them that every experience in our classes will be comfortable or ethically clear. Certainly the ethical commitments they bring to the classroom could be challenged, especially if they hold on too tightly to particular notions of consent and inclusivity or fail to recognize their assumptions about partners, which might be ill-informed. Yet when students experience the uncertainty of ethical gray areas or challenges to their own ethical commitments, the result need not be negative. Our contention is that such experiences are prime sites for learning critical skills required of all practitioners. We use several strategies in our teaching to turn these challenges into learning moments. For one, we use the classroom, especially experiential learning activities, to prepare students for the engaged elements of field-based courses, such as interviewing, running a focus group, and observing and analyzing.[9] Working through ethical dilemmas in a classroom setting, while challenging, is preferable to confronting them for the first time in the midst of partners, clients, and bystanders. Second, we try to be explicit with students about our intentions in positioning them within ethical dilemmas and in holding back from offering easy solutions. As a third strategy, our courses provide multiple opportunities for reflection, both peer to peer as well as student to instructor, so that students can find their ethical footing without prematurely abandoning their own ethical commitments. Fourth, we find that separating the most difficult ethical dilemmas from graded assignments allows students more flexibility to explore ethical commitments and strategies of acting upon them. Finally, field-based courses where the local partners and clients include students offer tremendous opportunities for our students to come to appreciate different perspectives held by people who are their peers on at least one important dimension. Experiential courses and exercises are among the richest learning experiences we have witnessed.[10] In the same spirit, coteaching and frequent consultations and debriefs with faculty colleagues offer us as instructors the opportunity to reflect on and reevaluate our own ethical commitments and aims as teacher-practitioners.

Notes

1. Susan F. Hirsch and Agnieszka Paczyńska, "Ethics and Field-Based Courses: How to Prepare Students for the Challenges of Practice," in *Conflict Zone, Comfort Zone: Ethics,*

Pedagogy, and Effecting Change in Field-Based Courses, ed. Agnieszka Paczyńska and Susan F. Hirsch (Athens: Ohio University Press, 2019), 21–40.

2. Hirsch and Paczyńska, "Ethics and Field-Based Courses."

3. Agnieszka Paczyńska and Susan F. Hirsch, eds., *Conflict Zone, Comfort Zone: Ethics, Pedagogy, and Effecting Change in Field-Based Courses* (Athens: Ohio University Press, 2019).

4. Hirsch and Paczyńska, "Ethics and Field-Based Courses."

5. Brian Leiter, "Academic Ethics: The Legal Tangle of 'Trigger Warnings,'" *Chronicle of Higher Education*, November 13, 2016, https://www.chronicle.com/article/academic-ethics-the-legal-tangle-of-trigger-warnings/.

6. Choi-Fitzpatrick, Irvin-Erickson, and Verdeja, introduction to this volume.

7. Leslie Dwyer and Alison Castel, "Framing 'Experience' in International Field-Based Learning," in Paczyńska and Hirsch, *Conflict Zone, Comfort Zone*, 41–63.

8. Hirsch and Paczyńska, "Ethics and Field-Based Courses."

9. Agnieszka Paczyńska, "Teaching Conflict and Conflict Resolution," in *Handbook in Political Science and International Relations*, ed. John Ishiyama, Will Miller, and Eszter Simon (Northampton, MA: Edward Elgar, 2015), 173–184; Paczyńska and Hirsch, *Conflict Zone, Comfort Zone*.

10. Susan F. Hirsch and Agnieszka Paczyńska, *Teaching Peace and Conflict Studies* (Northampton, MA: Edward Elgar, forthcoming).

Bibliography

Abeegunde, Balogun. *Afrikan Martial Arts: Discovering the Warrior Within*. Atlanta, GA: Boss Up, 2008.

Abrahamyan, Milena, Pervana Mammadova, and Sophio Tskhvariashvili. "Women Challenging Gender Norms and Patriarchal Values in Peacebuilding and Conflict Transformation across the South Caucasus." In "Challenging Gender Norms, Dealing with the Past, and Protecting the Environment: Community-Driven Conflict Transformation in the South Caucasus," ed. Philip Gamaghelyan, Sevil Huseynova, Maria Karapetyan, Pınar Sayan. Special issue of *Journal of Conflict Transformation* 3, no. 1 (2018): 46–71.

Abrams, Kathryn. "Lawyers and Social Change Lawbreaking: Confronting a Plural Bar." *University of Pittsburgh Law Review* 52, no. 4 (1991): 753–784.

ACRI, Situation Report: The State of Human Rights in Israel and the OPT 2014, p. 23, *available at*: https://law.acri.org.il/en/2014/12/10/sitrep2014/ (last visited Oct. 29, 2021).

Action Group for Syria. "Final Communiqué." June 30, 2012. https://peacemaker.un.org/sites/peacemaker.un.org/files/SY_120630_Final%20Communique%20of%20the%20Action%20Group%20for%20Syria.pdf.

Addams, Jane. *Newer Ideals of Peace*. London: Macmillan, 1906.

African Union. *African Union Transitional Justice Policy*. Addis Ababa: African Union, 2019.

African Union. "Report of the High Level Panel of the African Union for the Resolution of the Crisis in Côte d'Ivoire." PSC/AHG/2(CCLXV), March 10, 2011.

Aga, Cavid, and Onnik James Krikorian. "No Hope for NK Peace? When Activists and Journalists Become Combatants." Meydan.TV, May 16, 2016. https://d9mc3ts4czbpr.cloudfront.net/az/article/no-hope-for-nk-peace-when-activists-and-journalists-become-combatants/.

Alston, Phillip. *The Parlous State of Poverty Eradication*. Report to the United Nations Human Rights Council. New York: United Nations, July 2, 2020.

American Bar Association. *Model Code of Professional Responsibility*. 1969. https://www.americanbar.org/content/dam/aba/administrative/professional_responsibility/2011build/mod_code_prof_resp.pdf.

American Bar Association. *Model Rules of Professional Conduct*. 2000. Chicago, Illinois.

American Bar Association. "Model Rules of Professional Conduct: Table of Contents." 2020. https://www.americanbar.org/groups/professional_responsibility/publications/model_rules_of_professional_conduct/model_rules_of_professional_conduct_table_of_contents/.

American Bar Association. "When Is It OK for a Lawyer to Lie?" Chicago, IL. December 2018.

Anda, Robert F., Vincent J. Felitti, J. Douglas Bremner, John D. Walker, Charles Whitfield, Bruce D. Perry, Shanta R. Dube, and Wayne H. Giles. "The Enduring Effects of Abuse and Related Adverse Experiences in Childhood: A Convergence of Evidence from Neurobiology and Epidemiology." *European Archives of Psychiatry Clinical Neuroscience* 256, no. 3 (2006): 174–186. https://doi.org/10.1007/s00406-005-0624-4.

Anderson, Mary B. "Can My Good Intentions Make Things Worse? Lessons for Peacebuilding from the Field of International Humanitarian Aid." In *A Handbook of International Peacebuilding: Into the Eye of the Storm,* edited by John Paul Lederach and Janice Moomaw Jenner. San Francisco: Jossey-Bass, 2002: 225–234.

Anderson, Mary B. *Do No Harm: How Aid Can Support Peace—or War.* Boulder, CO: Lynne Rienner, 1999.

Anderson, Mary B., and Lara Olson. *Confronting War: Critical Lessons for Peace Practitioners.* Cambridge, MA: Collaborative for Development Action, Inc. Reflecting on Peace Practice Project, 2003.

Andreas, Peter. "Criminalizing Consequences of Sanctions: Embargo Busting and Its Legacy." *International Studies Quarterly* 49, no. 2 (June 2005): 335–360.

Ani, Marimba. *Yurugu: An African-Centered Critique of European Cultural Thought and Behavior.* Trenton, NJ: Africa World Press, 1994.

Annan, Kofi. "My Departing Advice on How to Save Syria." *Financial Times,* August 2, 2012. https://www.ft.com/content/b00b6ed4-dbc9-11e1-8d78-00144feab49a#ixzz22PUj6wxm.

Aptheker, Herbert. *American Negro Slave Revolts.* New York: International Publishing, 1983.

Apuuli, Kasaija. "The African Union's Notion of 'African Solutions to African Problems' and the Crises in Côte d'Ivoire (2010–2011) and Libya (2011)." *African Journal of Conflict Resolution* 12, no. 2 (2012): 135–160.

Armus, Teo. "After Helping Migrants in the Arizona Desert, an Activist Was Charged with a Felony. Now, He's Been Acquitted." *Washington Post,* November 21, 2019.

Arthur, Paige. "How Transitions Reshaped Human Rights: A Conceptual History of Transitional Justice." *Human Rights Quarterly* 31 (2009): 321–367.

Ashby, Muata. *Introduction to MA'AT Philosophy.* Lithonia, GA: Sema Institute, 2005.

Asikiwe, Jade. *Melanin: The Gift of the Cosmos.* Scotts Valley, CA: CreateSpace, 2018.

Autesserre, Severine. *Peaceland: Conflict Resolution and the Everyday Politics of International Intervention.* New York: Cambridge University Press, 2014.

Avent, Quashon. "African Americans and the 2nd Amendment: The Need for Black Armed Self-Defense." *The Carolinian,* February 20, 2019. https://carolinianuncg.com/2019/02/20/african-americans-and-the-2nd-amendment-the-need-for-black-armed-self-defense/.

Avruch, Kevin. "Does Our Field Have a Centre?" *International Journal of Conflict Engagement and Resolution* 1, no. 1 (2013): 10–31.

Babbitt, Eileen, and Ellen L. Lutz, eds. *Human Rights and Conflict Resolution in Context: Colombia, Sierra Leone, and Northern Ireland.* Syracuse, NY: Syracuse University Press, 2009.

Baker, General. "History of Struggle in Detroit." Lecture organized by the Poverty Initiative Poverty Scholars Program at the United States Social Forum, Detroit, MI, June 24, 2010.

Balagoon, Kuwasi. *A Soldier's Story: Revolutionary Writings by a New African Anarchist.* Montreal: Kersplebedeb, 2019.

Barber, William J., II. *The Third Reconstruction: Moral Mondays, Fusion Politics, and the Rise of a New Justice Movement.* Boston: Beacon Press, 2016.

Belgioioso, Margherita, Stefano Costalli, and Kristian Skrede Gleditsch. "Better the Devil You Know? How Fringe Terrorism Can Induce an Advantage for Moderate Nonviolent Campaigns." *Terrorism and Political Violence* 33, no. 3 (March 2019): 596–615.

Bell, Derrick. *And We Are Not Saved: The Elusive Quest for Racial Justice.* New York: Basic Books, 1987.

Bell, Derrick. *Faces at the Bottom of the Well*. New York: Basic Books, 1992.

Bell, Derrick. "A Holiday for Dr. King: The Significance of Symbols in the Black Freedom Struggle." *U.C. Davis Law Review* 17, no. 2 (1984): 433–444.

Bell, Derrick. "Racial Realism." *Connecticut Law Review* 24, no. 2 (1992): 363–379.

Bell, Derrick. "Racial Reflections: Dialogues in the Direction of Liberation." *UCLA Law Review* 37, no. 6 (1990): 1037–1100.

Bell, Derrick. "White Superiority in America: Its Legal Legacy, Its Economic Costs." *Villanova Law Review* 33, no. 5 (1988): 767–779.

Bellamy, Alex J. "The Responsibilities of Victory: *Jus Post Bellum* and the Just War." *Review of International Studies* 34, no. 4 (October 2008): 601–625.

Bellamy, Alex, and Paul Williams. "The New Politics of Protection? Côte d'Ivoire, Libya and the Responsibility to Protect." *International Affairs* 87, no. 4 (2011): 825–850.

Berlin, Isaiah. *Liberty*. New York: Oxford University Press, 2002.

Bersheid, Ellen. "Opinion Change and Communicator-Communicatee Similarity and Dissimilarity." *Journal of Personality and Social Psychology* 4, no. 6 (1966): 670–680.

Black Demographics. "Black Male Statistics: Population." Accessed August 22, 2020. http://blackdemographics.com/population/black-male-statistics/.

Black Liberation Army. *Message to the Black Movement: A Political Statement from the Black Underground*. Chico, CA: Abraham Guillen Press/Arm the Spirit, 1971.

Bloch, Nadine. *Education and Training in Nonviolent Resistance*. Washington, D.C.: U.S. Institute of Peace, 2016.

Booth, Ken. "Beyond Critical Security Studies." In *Critical Security Studies and World Politics*, edited by Ken Booth, 280–310. Boulder, CO: Lynne Rienner, 2005.

Boraine, Alexander, L. "Transitional Justice." In *Pieces of the Puzzle: Keywords on Reconciliation and Transitional Justice*, edited by Charles Villa-Vicencio and Erik Doxtader, 67–72. Cape Town: Institute for Justice and Reconciliation, 2004.

Boyce, Geoffrey Alan, Samuel N. Chambers, and Sarah Launius. "Bodily Inertia and the Weaponization of the Sonoran Desert in US Boundary Enforcement: A GIS Modeling of Migration Routes through Arizona's Altar Valley." *Journal on Migration and Human Security* 7, no. 1 (March 6, 2019). doi: https://doi.org/10.1177/23315 02419825610.

Bronner, Stephen Eric. *Ideas in Action: Political Tradition in the Twentieth Century*. Lanham, MD: Rowman & Littlefield, 2000.

brown, adrienne maree. *Emergent Strategy: Shaping Change, Changing Worlds*. Chico, CA: AK Press, 2017.

Buchanan-Smith, Margie. "How the Sphere Project Came into Being: A Case Study of Policy-Making in the Humanitarian Aid Sector and the Relative Influence of Research." Overseas Development Institute, July 1, 2003. https://www.odi.org/sites/odi.org. uk/files/odi-assets/publications-opinion-files/176.pdf.   https://cdn.odi.org/media/documents/176.pdf.

Burton, John. *Conflict: Resolution and Prevention*. New York: St. Martin's Press, 1990.

Burton, Mark, and Carolyn Kagan. "Liberation Social Psychology: Learning from Latin America." *Journal of Community & Applied Social Psychology* 15, no. 1 (2004): 63–78.

Case, Benjamin. "Riots as Civil Resistance: Rethinking the Dynamics of 'Nonviolent' Struggle." *Journal of Resistance Studies* 1, no. 4 (2018): 9–44.

CDA Collaborative Learning Projects. "Do No Harm Workshop Participants Manual." December 2016. https://www.cdacollaborative.org/publication/no-harm-workshop-participants-manual-2016/.

Charlton, James I. *Nothing about Us without Us: Disability Oppression and Empowerment*. Berkeley: University of California Press, 1998.

Checkel, Jeffrey. "The Constructivist Turn in International Relations Theory." *World Politics* 50, no. 2 (1998): 324–348.

Chenoweth, Erica, and Kurt Schock. "Do Contemporaneous Armed Challenges Affect the Outcomes of Mass Nonviolent Campaigns?" *Mobilization: An International Quarterly* 20, no. 4 (December 2015): 427–451.

Chenoweth, Erica, and Maria Stephan. *Why Civil Resistance Works: The Strategic Logic of Nonviolent Conflict.* New York: Columbia University Press, 2011.

Chimni, B. S. "Globalization, Humanitarianism and the Erosion of Refugee Protection." *Journal of Refugee Studies* 13, no. 3 (2000): 243–263.

Churchill, Ward. *Pacifism as Pathology: Reflections on the Role of Armed Struggle in North America.* Chico, CA: AK Press, 2007.

Clarke, John Henrik. *Christopher Columbus and the African Holocaust: Slavery and the Rise of European Capitalism.* New York: A & B Books, 1992.

Cobb, Charles. *This Nonviolent Stuff'll Get You Killed: How Guns Made the Civil Rights Movement Possible.* Durham, NC: Duke University Press, 2015.

Coggins, T. "Translating Wellbeing Policy, Theory and Evidence into Practice." In *Wellbeing, Recovery and Mental Health,* edited by Mike Slade, Lindsey Oades, and Aaron Jarden, 231–244. Cambridge: Cambridge University Press, 2017.

Cohen, Andrew. "The Torture Memos, 10 Years Later." *The Atlantic,* February 6, 2012.

Cohen, Seth B. 'The Challenging Dynamics of Global North-South Peacebuilding Partnerships: Practitioner Stories from the Field." *Journal of Peacebuilding & Development* 9, no. 3 (2014): 65–81. https://doi.org/10.1080/15423166.2014.984571.

Collins, William J., and Robert A. Margo. "The Economic Aftermath of the 1960s Riots in American Cities: Evidence from Property Values." *Journal of Economic History* 67, no. 4 (2007): 849–883.

Cook, Nicolas. "Côte d'Ivoire's Post-Election Crisis." CRS Research Report for Congress. Congressional Research Service, January 28, 2011. https://www.refworld.org/pdfid/4d58e5832.pdf.

Cortright, David, and George A. Lopez. "Are Sanctions Just? The Problematic Case of Iraq." *Journal of International Affairs* 52, no. 2 (Spring 1999): 735–755.

Coser, Lewis A. *The Functions of Social Conflict.* London: Routledge, 1956.

Cover, Robert M. "The Folktales of Justice: Tales of Jurisdiction." *Capital University Law Review* 14, no. 2 (1985): 179–203.

Cunningam, Jamein P., and Rob Gillezeau. "Don't Shoot! The Impact of Historical African American Protest on Police Killings of Civilians." *Journal of Quantitative Criminology* 37, no. 1 (2019): 1–34.

Curry, Tommy. *The Man-Not: Race, Class, Genre, and the Dilemmas of Black Manhood.* Philadelphia, PA: Temple University Press, 2017.

Davenport, Christian. *How Social Movements Die: Repression and Demobilization of the Republic of New Africa.* Boston: Cambridge University Press, 2014.

Davenport, Christian. "Multi-Dimensional Threat Perception and State Repression: An Inquiry into Why States Apply Negative Sanctions." *American Journal of Political Science* 39, no. 3 (1995): 683–713.

Della Porta, Donatella. *Clandestine Political Violence.* Cambridge: Cambridge University Press, 2013.

De Waal, Alex. "Intervention Unbound." *Index on Censorship* 6 (1994): 13–31.

de Zayas, Alfred. "Report of the Independent Expert on the Promotion of a Democratic and Equitable International Order on His Country Visit to the Bolivarian Republic of

Venezuela: Comments by the State." United Nations Human Rights Council, New York, NY, September 10, 2018.

Dilanyan, Sona, Aia Beraia, and Hilal Yavuz. "Beyond NGOs: Decolonizing Peacebuilding and Human Rights." In "Challenging Gender Norms, Dealing with the Past, and Protecting the Environment: Community-Driven Conflict Transformation in the South Caucasus," ed. Philip Gamaghelyan, , Sevil Huseynova, Maria Karapetyan, Pınar Sayan. Special issue of *Journal of Conflict Transformation* 3, no. 1 (2018): 157–173.

Drezner, Daniel W. "Sanctions Sometimes Smart: Targeted Sanctions in Theory and Practice." *International Studies Review* 13, no. 1 (March 2011): 96–108.

"Dr. King Warns That Riots Might Bring Rightists' Rule." *New York Times*, February 18, 1968.

Du Toit, Fanie. "A Double-Edged Sword." In *Reconciliation: A Guiding Vision for South Africa?*, edited by Ernst Conradie, 88–91. Johannesburg: Ecumenical Foundation of Southern Africa, 2013.

Dunbar-Ortiz, Roxanne. *An Indigenous Peoples' History of the United States*. Boston: Beacon Press, 2014.

Dunbar-Ortiz, Roxanne. *Loaded: A Disarming History of the Second Amendment*. San Francisco: City Lights, 2018.

Dwyer, Leslie, and Alison Castel. "Framing 'Experience' in International Field-Based Learning." In *Conflict Zone, Comfort Zone: Ethics, Pedagogy, and Effecting Change in Field-Based Courses*, edited by Agnieszka Paczyńska and Susan F. Hirsch, 41–63. Athens: Ohio University Press, 2019.

Eagly, Alice H., Wendy Wood, and Shelly Chaiken. "Causal Inferences about Communicators and Their Effect on Attitude Change." *Journal of Personality and Social Psychology* 36, no. 4 (1978): 424–435.

Early, Bryan R. *Busted Sanctions: Explaining Why Economic Sanctions Fail*. Stanford, CA: Stanford University Press, 2015.

Early, Bryan R., and Marcus Schulzke. "Still Unjust in Different Ways: How Targeted Sanctions Fall Short of Just War Theory's Principles." *International Studies Review* 21, no. 1 (April 2018): 123–142.

Easthom, Tiffany. "Unarmed Approaches to Civilian Protection." Statement given at International Peace Institute, New York, September 15, 2015. https://www.youtube.com/watch?v=vd3bIOFY0Fk.

Economic Community of West African States. "Final Communiqué, Extraordinary Session of the Authority of Heads of State and Government on Côte d'Ivoire." Press release 192/2010. December 24, 2010.

English, Michael D. *The US Institute of Peace: A Critical History*. Boulder, CO: Lynne Rienner, 2018.

European Commission. *Ethics for Researchers*. Luxembourg: European Commission, 2013.

Economic Community of West African States, "Resolution A/RES.1/03/11 of the Authority of Heads of State and Government on the Situation in Côte d'Ivoire," press release 043/2011, March 25, 2011.

European Parliament and Council of the European Union. General Data Protection Regulation. Law 119. 2018. https://ec.europa.eu/info/law/law-topic/data-protection_en.

European Union. "Consolidated Versions of the Treaty on European Union and the Treaty on the Functioning of the European Union." Publications Office of the European Union, 2016. https://www.consilium.europa.eu/media/29726/qc0116985enn.pdf.

Fanon, Franz. *The Wretched of the Earth*. London: Penguin Books, 1967.

Fast, Larissa. *Aid in Danger: The Perils and Promise of Humanitarianism*. Philadelphia: University of Pennsylvania Press, 2014.

Feliz, Wendy. "Providing Humanitarian Aid to People Crossing the Border Should Never Be a Crime." *Immigration Impact*. June 17, 2019.

Finnemore, Martha, and Kathryn Sikkink. "International Norm Dynamics and Political Change." *International Organization* 52, no. 4 (1998): 887–917.

Firchow, Pamina. *Reclaiming Everyday Peace: Local Voices in Measurement and Evaluation after War*. Cambridge: Cambridge University Press, 2018.

Focus Economics. "Venezuela Economic Outlook." In *Focus Economics: Economic Forecasts from the World's Leading Economists*. InSight Crime, July 14, 2020. https://www.focus-economics.com/countries/venezuela.

Fontan, Victoria. *Descolonización de la Paz*. Cali: Pontificia Universidad Sello Editorial Javeriano, 2013.

Freedman, Monroe H., and Abbe Smith. *Understanding Lawyers' Ethics*. 3rd ed. Newark, NJ: LexisNexis, 2004.

Freire, Paulo. *Pedagogy of the Oppressed*. New York: Herder and Herder, 1970.

Fukuyama, Francis. *The End of History and the Last Man*. New York: Free Press, 2006.

Galera, Sandra, Melissa Tracy, Katherine J. Hoggatt, Charles DiMaggio, and Adam Karpati. "Estimated Deaths Attributable to Social Factors in the United States." *American Journal of Public Health* 101, no. 8 (2011): 1456–1465.

Galtung, Johan. "Violence, Peace, and Peace Research." *Journal of Peace Research* 6, no. 3 (January 1969): 167–191.

Gamaghelyan, Philip. *Conflict Resolution beyond the International Relations Paradigm: Evolving Designs as Transformative Practices in Nagorno-Karabakh and Syria*. Soviet and Post-Soviet Politics and Societies Series. Stuttgart: ibidem Press/Columbia University Press, 2017.

Gamaghelyan, Philip, Sevil Huseynova, Christina Soloyan, Pınar Sayan, Sophio Tskvariashvili, eds. "Activism and Peace in the South Caucasus." *Caucasus Edition* 4, no. 2 (2019): 1–155.

Gamaghelyan, Philip, Sevil Huseynova, Christina Soloyan, Pınar Sayan, Sophio Tskvariashvili, eds. "Exploring Post-Liberal Approaches to Peace in the South Caucasus." *Caucasus Edition* 4, no. 1 (November 2019): 1–186.

Gaskew, Tony. *Stop Trying to Fix Policing: Lessons Learned from the Front Lines of Black Liberation*. Lanham, MD: Rowman & Littlefield, 2021.

Gelderloos, Peter. *How Nonviolence Protects the State*. Boston: Detritus Books, 2018.

Gheciu, Alexandra, and Jennifer Welsh. "The Imperative to Rebuild: Assessing the Normative Case for Postconflict Reconstruction." *Ethics & International Affairs* 23, no. 2 (Summer 2009): 121–146.

Giddens, Anthony. *New Rules of Sociological Method: A Positive Critique of Interpretative Sociologies*. Stanford, CA: Stanford University Press, 1993.

Goetze, Catherine. *The Distinction of Peace: A Social Analysis of Peacebuilding*. Ann Arbor: University of Michigan Press, 2017.

Goines, Beverly Janet. "Social Justice as Peacebuilding in Black Churches: Where Do We Go from Here?" In *Building Peace in America*, edited by Emily Sample and Douglas Irvin-Erickson, 23–38. Lanham, MD: Rowman and Littlefield, 2020.

Goodman, Amy, and Denis Moynahan. "Humanitarian Groups Continue Work at U.S.-Mexico Border, Despite Risk of Prosecution." *Rabble.ca*, August 22, 2019. https://rabble.ca/columnists/2019/08/humanitarian-groups-continue-work-us-mexico-border-despite-risk-prosecution.

Gordon, Joy. "A Peaceful, Silent, Deadly Remedy: The Ethics of Economic Sanctions." *Ethics & International Affairs* 13, no. 1 (April 2006): 123–142.

Gordon, Joy. "Roundtable: Economic Sanctions and Their Consequences." *Ethics & International Affairs* 33, no. 3 (September 2019): 275–302.

Gordon, Joy. "Smart Sanctions Revisited." *Ethics & International Affairs* 25, no. 3 (Fall 2011): 315–335.

Green, Helen Taylor, Shaun Gabbidon, and Sean Wilson. "Included? The Status of African American Scholars in the Discipline of Criminology and Criminal Justice Since 2004." *Journal of Criminal Justice Education* 29, no. 1 (2018): 96–115.

Grono, Nick, and Adam O'Brien. "Justice in Conflict? The ICC and Peace Processes." In *Courting Conflict? Justice, Peace and the ICC in Africa*, edited by Nicholas Waddell and Phil Clark, 13–20. London: Royal African Society, 2008.

Guardian Staff. "Civil Rights Heroine Claudette Colvin." *The Guardian*, December 15, 2000. https://www.theguardian.com/theguardian/2000/dec/16/weekend7.weekend12.

Haines, Herbert H. *Black Radicals and the Civil Rights Mainstream, 1954–1970*. Knoxville, TN: University of Tennessee Press, 1995.

Hamby, Sherry, Lawrence Grandpre, John Hope Bryant, and Noche Diaz. "When Is Rioting the Answer?" *Zócalo*, July 8, 2015. https://www.zocalopublicsquare.org/2015/07/08/when-is-rioting-the-answer/ideas/up-for-discussion/.

Harding, Sandra. "Rethinking Standpoint Epistemology: What Is 'Strong Objectivity'?" *Centennial Review* 36, no. 3 (Fall 1992): 437–470.

Hauss, Charles. *Security 2.0: Dealing with Global Wicked Problems*. Lanham, MD: Rowman & Littlefield, 2015.

High-Level Panel on Peace Operations. *Uniting Our Strengths for Peace*. New York: United Nations, June 2015.

Hill, Tom. "Kofi Annan's Multilateral Strategy of Mediation and the Syrian Crisis: The Future of Peacemaking in a Multipolar World?" *International Negotiation* 20, no. 3 (2015): 444–478.

Hilliard, Asa, Larry Williams, and Nia Damali. *The Teachings of Ptahhotep: The Oldest Book in the World*. Scotts Valley, CA: CreateSpace, 1987.

Hinnebusch, Raymond, and I. William Zartman with Elizabeth Parker-Magyar and Omar Imady. *UN Mediation in the Syrian Crisis: From Kofi Annan to Lakhdar Brahimi*. New York: International Peace Institute, 2016.

Hirsch, Susan F., and Agnieszka Paczyńska. "Ethics and Field-Based Courses: How to Prepare Students for the Challenges of Practice." In *Conflict Zone, Comfort Zone: Ethics, Pedagogy, and Effecting Change in Field-Based Courses*, edited by Agnieszka Paczyńska and Susan F. Hirsch, 21–40. Athens: Ohio University Press, 2019.

Hirsch, Susan F., Agnieszka Paczyńska. *Teaching Peace and Conflict Studies*. Northampton, MA: Edward Elgar, forthcoming.

Humane Borders/Fronteras Compasivas. "Migrant Death Mapping, Migrants Deaths, Rescue Beacons, Water Stations 2000–2018." Accessed July 23, 2020. https://humane borders.org/migrant-death-mapping/.

Hume, Liz, and Conor Seyle. "COVID-19 Is Stabilization in Reverse: Applying the Principles of the Global Fragility Act to Pandemic Response." Alliance for Peacebuilding, May 8, 2020. https://medium.com/@AfPeacebuilding/covid-19-is-stabilization-in-reve rse-applying-the-principles-of-the-global-fragility-act-to-6bd3baba4878.

Ienca, Marcello, Agata Ferretti, Samia Hurst, Milo Puhan, Christian Lovis, and Effy Vayena. "Considerations for Ethics Review of Big Data Health Research: A Scoping Review." *PLOS One* 13, no. 10 (October 2018). https://doi.org/10.1371/journal.pone.0204937.

Illich, Ivan, ed. *Disabling Professions.* Ideas in Progress series. New York: M. Boyars, 1987.

International Crisis Group. "Libya: Achieving a Ceasefire, Moving toward a Legitimate Government." Statement. May 13, 2011. https://www.crisisgroup.org/middle-east-north-africa/north-africa/libya/libya-achieving-ceasefire-moving-toward-legitimate-government.

International Crisis Group. "Popular Protest in North Africa and the Middle East (V): Making Sense of Libya." *Middle East/North Africa Report,* no. 107 (2011). https://www.crisisgroup.org/middle-east-north-africa/north-africa/libya/popular-protest-north-africa-and-middle-east-v-making-sense-libya.

International Committee of the Red Cross, Code of Conduct for the International Red Cross and Red Crescent Movement and Non-Governmental Organizations (NGOs) in Disaster Relief, December 12, 1994, https://www.icrc.org/en/doc/resources/documents/publication/p1067.htm.

International Federation of Red Cross and Red Crescent Societies, The Seven Fundamental Principles of the Red Cross and Red Crescent Societies, https://www.icrc.org/sites/default/files/topic/file_plus_list/4046-the_fundamental_principles_of_the_international_red_cross_and_red_crescent_movement.pdf.

Jameson, Fredric. *Valences of the Dialectic.* London: Verso, 2009.

Jemisin, N. K. *How Long 'til Black Future Month?* New York: Orbit, 2018.

Jenkins, J. Craig, and Charles Perrow. "Insurgency of the Powerless: Farm Worker Movements (1946–1972)." *American Sociological Review* 42, no. 2 (1977): 249–268.

Johnson, Nicholas. *Negroes and the Gun: The Black Tradition of Arms.* Buffalo, NY: Prometheus Books, 2014.

Joinet, Louis. "Final Report on the Administration of Justice and the Question of Impunity" United Nations Committee on Human Rights (New York: United Nations, 1997). http://hrlibrary.umn.edu/demo/RightsofDetainees_Joinet.pdf.

Joseph, Peniel E. *The Sword and the Shield: The Revolutionary Lives of Malcolm X and Martin Luther King Jr.* New York: Basic Books, 2020.

Kadivar, Mohammad Ali, and Neil Ketchley. "Sticks, Stones, and Molotov Cocktails: Unarmed Collective Violence and Democratization." *Socius* 4 (January 2018): 1–16.

Kamene, Kaba Hiawatha. *Spirituality before Religions: Spirituality Is Unseen Science.* Scotts Valley, CA: CreateSpace, 2019.

Karade, Baba Ifa. *The Handbook of Yoruba Religious Concepts.* New York: Weiser Books, 1994.

Karanth, Sanjana. "Federal Prosecutors to Retry Border Humanitarian Volunteer for Helping Migrants." *Huffington Post,* July 2, 2019.

Karlin, Mara. "After 7 Years of War, Assad Has Won in Syria. What's Next for Washington?" Brookings Institute, February 13, 2018. https://www.brookings.edu/blog/order-from-chaos/2018/02/13/after-7-years-of-war-assad-has-won-in-syria-whats-next-for-washington/.

Kelman, Herbert C., and Donald P. Warwick. "The Ethics of Social Intervention: Goals, Means, and Consequences." In *The Ethics of Social Intervention,* edited by Gordon Bermant, Herbert C. Kelman, and Donald P. Warwick, 3–36. Washington, D.C.: Hemisphere, 1978.

Kester, Kevin, Michalinos Zembylas, Loughlin Sweeney, Kris Hyesoo Lee, Soonjung Kwon, and Jeongim Kwon. "Reflections on Decolonizing Peace Education in Korea: A Critique and Some Decolonial Pedagogic Strategies." *Teaching in Higher Education,* no. 2 (2019): 145–164. https://doi.org/10.1080/13562517.2019.1644618.

Khader, Serene. *Decolonizing Universalism: A Transnational Feminist Ethic.* New York: Oxford University Press, 2019.

Khagram, Sanjeev, James Riker, and Kathryn Sikkink. "From Santiago to Seattle: Transnational Advocacy Groups and Restructuring World Politics." In *Restructuring World Politics: Transnational Social Movements, Networks, and Norms,* edited by Sanjeev Khagram, James Riker, and Kathryn Sikkink, 3–23. Minneapolis: University of Minnesota Press, 2002.

King, Martin Luther, Jr. "Acceptance Address for the Nobel Peace Prize." Oslo, December 10, 1964. Martin Luther King, Jr. Research and Education Institute, Stanford University, https://kinginstitute.stanford.edu/.

King, Martin Luther, Jr. "The Other America." In *The Radical King,* edited by Cornell West, 235–245. United States: Beacon Press, 2015.

King, Martin Luther, Jr. *Stride toward Freedom: The Montgomery Story.* Boston: Beacon Press, 1958.

King, Martin Luther, Jr. *Trumpet of Conscience.* New York: Harper and Row, 1967.

King, Martin Luther, Jr. *Why We Can't Wait.* 1964; New York: Signet Classics, 2000.

King, Martin Luther, Jr. "Beyond Vietnam." In *A Time to Break the Silence: The Essential Works of Martin Luther King, Jr.,* edited by Walter Dean Myers. Boston: Beacon Press, 2013.

Klotz, Audie. "Transnational Activism and Global Transformations: The Anti-Apartheid and Abolitionist Experiences." *European Journal of International Relations* 8, no. 1 (2002): 49–76.

Konigsberg v. State Bar of California, 353 U.S. 252 (1957).

Kriesberg, Louis. "The Development of the Conflict Resolution Field." In *Peacemaking in International Conflict: Methods and Techniques,* edited by I. William Zartman and J. Lewis Rasmussen, 51–77. Washington D.C.: U.S. Institute of Peace Press, 1997.

Laue, James, and Gerald Cormick. "The Ethics of Intervention in Community Disputes." In *Ethics of Social Intervention,* edited by Gordon Bermant, Herbert C. Kelman, and Donald P. Warwick. Washington, D.C.: Hemisphere, 1978.

Laya, Patricia, Ethan Bronner, and Tim Ross. "Maduro Stymied in Bid to Pull $1.2 Billion of Gold from U.K." *Bloomberg News,* January 25, 2019. https://www.bloomberg.com/news/articles/2019-01-25/u-k-said-to-deny-maduro-s-bid-to-pull-1-2-billion-of-gold.

Lazarus, Ned. "Evaluating Peace Education in the Oslo-Intifada Generation: A Long-Term Impact Study of Seeds of Peace 1993–2010." PhD dissertation, American University, 2011.

Lederach, John Paul. *Preparing for Peace: Conflict Transformation across Cultures.* Syracuse, NY: Syracuse University Press, 1995.

Le Guin, Ursula K. *The Ones Who Walk Away from Omelas.* Mankato, MN: Creative Education, 1993.

Leiter, Brian. "Academic Ethics: The Legal Tangle of 'Trigger Warnings.'" *Chronicle of Higher Education,* November 13, 2016. https://www.chronicle.com/article/academic-ethics-the-legal-tangle-of-trigger-warnings/.

Little, David. "Religion, Peace and the Origins of Nationalism." In *The Oxford Handbook of Religion, Conflict, and Peacebuilding,* edited by R. Scott Appleby, Atalia Omer, and David Little, 70–104. Oxford: Oxford University Press, 2015.

Logan, Manseen. "3 Speeches from Coretta Scott King That Commemorate MLK and Cement Their Family's Lasting Legacy." *Blavity,* January 23, 2020. https://blavity.com/

blavity-original/3-speeches-from-coretta-scott-king-that-commemorate-mlk-and-cement-their-familys-lasting-legacy?category1=culture.

Lopez, George A. "Targeted Sanctions." In *Combating Criminalized Power Structures: A Toolkit*, edited by Michael Dziedzic, 61–70. Lanham, MD: Rowman & Littlefield, 2016.

Lorde, Audre. *Sister/Outsider: Essays and Speeches*. Freedom, CA: Crossing Press, 1984.

Luban, David. "The Conscience of a Prosecutor." *Valparaiso University Law Review* 45, no. 1 (2010): 1–31.

Luban, David. "The Defense of Torture." Review of *War by Other Means: An Insider's Account of the War on Terror*, by John Yoo. *New York Review of Books*, March 15, 2007.

Luban, David. "Legal Ideals and Moral Obligations: A Comment on Simon." *William & Mary Law Review* 38, no. 1 (1996): 255–267.

Lynch, Colum. "On Ivory Coast Diplomacy, South Africa Goes Its Own Way." *Foreign Policy*, February 23, 2011. http://foreignpolicy.com/2011/02/23/on-ivory-coast-diplomacy-south-africa-goes-its-own-way/.

Mac Ginty, Roger. *International Peacebuilding and Local Resistance: Hybrid Forms of Peace*. London: Palgrave, 2011.

Malcolm X. *By Any Means Necessary: Malcolm X's Speeches and Writings*. New York: Pathfinder Press, 1992.

Malcolm X. "The Ballot or the Bullet." In *Malcolm X Speaks: Selected Speeches and Statements*, edited by George Breitman, 20–37. New York: Grove.

Mamdani, Mahmood. *Saviors and Survivors: Darfur, Politics, and the War on Terror*. New York: Doubleday, 2010.

Mangan, Fiona. "Illicit Drug Trafficking and Use in Libya: Highs and Lows." Peaceworks, May 2020. https://www.usip.org/publications/2020/05/illicit-drug-trafficking-and-use-libya-highs-and-lows.

Matalon, Lorne. "Extending 'Zero Tolerance' to People Who Help Migrants along the Border." *All Things Considered*, National Public Radio, May 28, 2019. https://www.npr.org/2019/05/28/725716169/extending-zero-tolerance-to-people-who-help-migrants-along-the-border.

Matsuda, Mari. "When the First Quail Calls: Multiple Consciousness as Jurisprudential Method." *Women's Rights Law Reporter* 11, nos. 1 (1989): 7–10.

Mayer, Jeremy D. "Nixon Rides the Backlash to Victory: Racial Politics in the 1968 Presidential Campaign." *The Historian* 64, no. 2 (December 2001): 351–366.

"Mbeki: Côte d'Ivoire Rivals Must Talk to End the Crisis." *Mail and Guardian*, January 24, 2011. http://mg.co.za/article/2011-01-24-mbeki-cocircte-divoire-rivals-must-talk-to-end-crisis.

McAdam, Doug. *Freedom Summer*. New York: Oxford University Press, 1988.

McAdam, Doug. *Political Process and the Development of Black Insurgency, 1930–1970*. Chicago: University of Chicago Press, 1982.

McCarthy, John D., and Mayer N. Zald. *The Trend of Social Movements in America: Resource Mobilization and Professionalization*. Morristown, NJ: General Learning Press, 1973.

Mearsheimer, John J. *The Tragedy of Great Power Politics*. New York: Norton, 2003.

Médicos por la Salud and República Bolivariana de Venezuela Asamblea Nacional. "Encuesta Nacional de Hospitales 2018." March 2019. https://cifrasonlinecomve.files.wordpress.com/2018/03/enh-final_2018fin.pdf

Mertus, Julie, and Jeffrey W. Helsing, eds. *Human Rights and Conflict: Exploring the Links between Rights, Law, and Peacebuilding*. Washington, D.C.: U.S. Institute of Peace Press, 2006:

Michael, Kobi, and Eyal Ben-Ari. "Contemporary Peace Support Operations: The Primacy of the Military and Internal Contradictions." *Armed Forces and Society* 37, no. 4 (2011): 657–679.

Miller, Darrell. "Retail Rebellion and the Second Amendment." *Indiana Law Journal* 86, no. 3 (2011): 939–976.

Mills, C. Wright. *The Sociological Imagination.* 1959; New York: Oxford University Press, 2000.

Minow, Martha. "Breaking the Law: Lawyers and Clients in Struggles for Social Change." *University of Pittsburgh Law Review* 52, no. 4 (1991): 723–752.

Montoya-Galvez, Camilo. "Think of Me as the Grim Reaper: McConnell Vows to Thwart Democratic Proposals." CBS News, April 22, 2019. https://www.cbsnews.com/news/mitch-mcconnell-vows-to-be-the-grim-reaper-to-thwart-all-democratic-proposals/.

Muñoz, Jordi, and Eva Anduiza. "'If a Fight Starts, Watch the Crowd': The Effect of Violence on Popular Support for Social Movements." *Journal of Peace Research* 56, no. 4 (2019): 485–498.

Muntaqim, Jalil. *On the Black Liberation Army.* Montreal: Montreal Anarchist Black Cross, 1997.

Muntaqim, Jalil. *We Are Our Own Liberators.* Chicago: Arissa, 2010.

Murithi, Tim. *The Ethics of Peacebuilding.* Edinburgh: Edinburgh University Press, 2009.

Myers, Daniel J. "Ally Identity: The Politically Gay." In *Identity Work in Social Movements,* edited by Jo Reger, Daniel J. Myers, and Rachel Einwohner, 167–187. Minneapolis: University of Minnesota Press, 2008.

Nadelman, Ethan. "Global Prohibition Regimes: The Evolution of Norms in International Society." *International Organization* 44, no. 4 (1990): 479–526.

Nader, Laura. *Harmony Ideology: Justice and Control in a Zapotec Mountain Village.* Redwood City, CA: Stanford University Press, 1991.

Naím, Moisés. "Democracy's Dangerous Impostors." *Washington Post,* April 21, 2007. http://www.washingtonpost.com/wp-dyn/content/article/2007/04/20/AR2007042001594.html.

Nambiar, Vijay. "Dear President Mbeki: The United Nations Helped Save the Ivory Coast." *Foreign Policy,* August 17, 2011. http://foreignpolicy.com/2011/08/17/dear-president-mbeki-the-united-nations-helped-save-the-ivory-coast/.

Nathan, Laurie. "The International Peacemaking Dilemma: Ousting or Including the Villains?" *Swiss Political Science Review* 26, no. 5 (2020): 468–486.

Nathan, Laurie. "No Ownership, No Peace: The Darfur Peace Agreement." Working Paper 5. Crisis States Research Centre, London School of Economics, 2006.

National African-American Gun Association. Accessed August 22, 2020. https://naaga.co/.

National Institutes of Health, Office of Intramural Research. "Dual-Use Research." September 12, 2020. https://oir.nih.gov/sourcebook/ethical-conduct/special-research-considerations/dual-use-research.

National Union of the Homeless. *National Organizing Drive of the National Union of the Homeless.* Pamphlet from the personal archive of Willie Baptist, 1988.

Neufeldt, Reina C. *Ethics for Peacebuilders: A Practical Guide.* Lanham, MD: Rowman and Littlefield, 2016.

Neufeldt, Reina C. "When Good Intentions Are Not Enough." *Conrad Grebel Review* 36, no. 2 (2018): 114–132.

Newman, Edward, Roland Paris, and Oliver Richmond, eds. *New Perspectives on Liberal Peacebuilding.* Tokyo: United Nations University Press, 2009.

NGO Monitor. "Facilitating Dialogue: EU-Funded NGOs and the Nagorno-Karabakh Conflict." October 2018. https://www.ngo-monitor.org/reports/facilitating-dialogue-eu-funded-ngos-and-the-nagorno-karabakh-conflict/.

Nkrumah, Kwame. *Handbook of Revolutionary Warfare*. Bedford, UK: Panaf Books, 1968.

Oberschall, Anthony. *Social Conflict and Social Movements*. Englewood Cliffs, NJ: Prentice-Hall, 1973.

Obi Desch, T. J. *Fighting for Honor: The History of African Martial Art in the Atlantic World*. Columbia: University of South Carolina Press, 2008.

Oldenhuis, Huibert. *Unarmed Civilian Protection*. Geneva: United Nations Institute for Training and Research and Nonviolent Peaceforce, 2016.

Omer, Atalia. "Decolonizing Religion and the Practice of Peace: Two Case Studies from the Postcolonial World." *Critical Research on Religion*, no. 3 (2020): 273–296. https://doi.org/10.1177/2050303220924111.

Orend, Brian. "Justice after War." *Ethics & International Affairs* 16, no. 1 (March 2002): 43–56.

Organisation for Economic Co-operation and Development. *States of Fragility 2018*. Paris: OECD Publishing, 2019. https://doi.org/10.1787/9789264302075-en.

Ortega, Bob. "Trial Begins for No More Deaths Volunteer Who Aided Migrants." *CNN Investigates*, June 3, 2019.

Paczyńska, Agnieszka. "Teaching Conflict and Conflict Resolution." In *Handbook in Political Science and International Relations*, edited by John Ishiyama, Will Miller, and Eszter Simon. Northampton, MA: Edward Elgar, 2015: 173–184.

Paczyńska, Agnieszka, and Susan F. Hirsch, eds. *Conflict Zone, Comfort Zone: Ethics, Pedagogy, and Effecting Change in Field-Based Courses*. Athens: Ohio University Press, 2019.

Palumbos, Robert M. "Within Each Lawyer's Conscience a Touchstone: Law, Morality and Attorney Civil Disobedience." *University of Pennsylvania Law Review* 153 (2005): 1057–1096.

Paris, Roland. "International Peacebuilding and the 'Mission Civilisatrice.'" *Review of International Studies* 28, no. 4 (2002): 637–656.

Parker, Kim, Juliana Menasce Horowitz, and Monica Anderson. "Majorities across Racial, Ethnic Groups Express Support for the Black Lives Matter Movement." Pew Research Center, Social & Demographic Trends Project, June 12, 2020. https://www.pewsocialtrends.org/2020/06/12/amid-protests-majorities-across-racial-and-ethnic-groups-express-support-for-the-black-lives-matter-movement/.

Peacebuilding Evaluation Consortium. "Guiding Steps for Peacebuilding Design, Monitoring, and Evaluation." 2018. http://www.dmeforpeace.org/wp-content/uploads/2018/06/PEC_Guiding-Steps-for-Peacebuilding-DME.pdf.

Peace Direct. *Atrocity Prevention and Peacebuilding*. 2018. https://www.peacedirect.org/us/wp-content/uploads/sites/2/2018/04/Atrocity-Prevention-Report_PD.pdf.

Pfeffer, Jeffrey, and Gerald R. Salancik. *The External Control of Organizations: A Resource Dependence Perspective*. Stanford Business Classics. 1978; Stanford, CA: Stanford Business Books, 2003.

Pinckney, Jonathan. *Making or Breaking Nonviolent Discipline in Civil Resistance Movements*. ICNC Monograph Series. Washington, D.C.: International Center on Nonviolent Conflict, 2016.

Poor People's Campaign. "Who Is Left Out of the COVID-19 Legislation." April 2020. https://www.poorpeoplescampaign.org/wp-content/uploads/2020/04/Who-is-left-out.pdf.

Price, Richard. *Maroon Societies: Rebel Slave Communities in the Americas*. Baltimore, MD: Johns Hopkins University Press, 1996.

Protecting Human Research Participants. "Protecting Human Research Participants: PHRP Training." https://phrptraining.com/. Last Accessed October 26, 2021.

Reuters Staff. "Text of Annan's Six-Point Peace Plan for Syria." Reuters, April 4, 2012. https://www.reuters.com/article/us-syria-ceasefire/text-of-annans-six-point-peace-plan-for-syria-idUSBRE8330HJ20120404.

Richmond, Oliver P. "Resistance and the Post-Liberal Peace." *Millennium: Journal of International Studies* 38, no. 3 (May 2010): 665–692.

Richmond, Oliver. *The Transformation of Peace: Rethinking Peace and Conflict Studies.* Basingstoke: Palgrave Macmillan, 2007.

Romashov, Vadim, Nuriyya Guliyeva, Lana Kokaia, and Tatia Kalatozishvili. "A Communitarian Peace Agenda for the South Caucasus: Supporting Everyday Peace Practices." In "Challenging Gender Norms, Dealing with the Past, and Protecting the Environment: Community-Driven Conflict Transformation in the South Caucasus," ed. Philip Gamaghelyan, Sevil Huseynova, Maria Karapetyan, and Pınar Sayan. *Journal of Conflict Transformation* 3, no. 1 (2018): 157–173.

Rosenthal, Naomi B. "Consciousness Raising: From Revolution to Re-Evaluation." *Psychology Women Quarterly* 8, no. 4 (June 1984): 309–326.

Roy, Sara. *Failing Peace, Gaza and the Palestinian-Israeli Conflict.* London: Pluto Press, 2007.

Roy, Sara. *Hamas and Civil Society in Gaza, Engaging the Islamist Social Sector.* 2nd ed. Princeton, NJ: Princeton University Press, 2014.

Rubenstein, Richard E. *Resolving Structural Conflicts: How Violent Systems Can Be Transformed.* Routledge Studies in Peace and Conflict Resolution. New York: Taylor & Francis, 2017.

Saad, Lydia. "What Percentage of Americans Own Guns." Gallup. Accessed August 22, 2020. https://news.gallup.com/poll/264932/percentage-americans-own-guns.aspx.

Sabaratnam, Meera. "History Repeating? Colonial, Socialist and Liberal Statebuilding in Mozambique." In *Routledge Handbook of International Statebuilding,* edited by David Chandler and Timothy D. Sisk, 106–117. London: Routledge, 2013.

Sanders, Kindaka. "A Reason to Resist: The Use of Deadly Force in Aiding Victims of Unlawful Police Aggression." *San Diego Law Review* 52, no. 3 (2015): 695–750.

Sandole, Dennis, and Ingrid Sandole-Staroste, eds. *Conflict Management and Problem Solving: Interpersonal to International Applications.* New York: New York University Press, 1987.

Sankara, Thomas. *Thomas Sankara Speaks: The Burkina Faso Revolution 1983–1987.* Atlanta, GA: Pathfinder, 1988.

Sarat, Austin, and Stuart Scheingold. *Cause Lawyering: Political Commitments and Professional Responsibilities.* Oxford: Oxford University Press, 1998.

Sarkar, Saurav, Shailly Gupta Barnes, Aaron Noffke, Sarah Anderson, Marc Bayard, Phyllis Bennis, John Cavanagh, Karen Dolan, Lindsay Koshgarian, Sam Pizzigati, Basav Sen, and Ebony Slaughter-Johnson. *The Souls of Poor Folk: Auditing America 50 Years after the Poor People's Campaign Challenged Racism, Poverty, the War Economy/Militarism and Our National Morality.* Institute for Policy Studies, 2018.

Sartre, Jean-Paul. "Preface to Frantz Fanon's Wretched of the Earth." In *The Wretched of the Earth* by Frantz Fanon, translated by Richard Philcox, xliii–lxii. New York: Grove Press, 2004.

Schirch, Lisa. *Ritual and Symbol in Peacebuilding.* Bloomfield, CT: Kumarian Press, 2005.

Schirch, Lisa. "10 Ethics of Peacebuilding." *Lisa Schirch* (blog), May 17, 2018. https://lisa schirch.wordpress.com/2018/05/17/10-ethics-of-peacebuilding/.

The Sentencing Project. "Race and Justice News: One-Third of Black Men Have Felony." October 10, 2017. https://www.sentencingproject.org/news/5593/.

Shumate, Michelle, and Lori Dewitt. "The North/South Divide in NGO Hyperlink Networks." *Journal of Computer-Mediated Communication* 13, no. 2 (2008): 405–428. https://doi.org/10.1111/j.1083-6101.2008.00402.x.

Siedlak, Monique Joiner. *Seven African Powers: The Orishas.* New York: Oshun, 2016.

Slim, Hugo. *Humanitarian Ethics: A Guide to the Morality of Aid in War and Disaster.* Oxford: Oxford University Press, 2015.

The Sphere. https://spherestandards.org

Sphere, Humanitarian Charter, accessed September 17, 2021, https://spherestandards. org/humanitarian-standards/humanitarian-charter/

Stratton, Allegra. "Obama, Cameron and Sarkozy: No Let Up in Libya until Gaddafi Departs." *The Guardian,* April 14, 2011. https://www.theguardian.com/world/2011/ apr/15/obama-sarkozy-cameron-libya.

Supreme Court of the United States. *Official Reports of the Supreme Court.* Vol. 561. Part 1. Washington, D.C.: U.S. Government Publication Office, 2010.

SURES: Estudios y Defensa en Derechos Humanos. "Unilateral Coercive Measures and Their Impact on the Human Rights of the Venezuelan People." Caracas. August 2018.

Tadjbakhsh, Shahrbanou, ed. *Rethinking the Liberal Peace: External Models and Local Alternatives.* London: Routledge, 2011.

Tarrow, Sidney. *Power in Movement: Social Movements, Collective Action, and Politics.* New York: Cambridge University Press, 1994.

Tarrow, Sidney. "States and Opportunities: The Political Structuring of Social Movements." In *Comparative Perspectives on Social Movements: Political Opportunities, Mobilizing Structures, and Cultural Framings,* edited by Douglas McAdam, John D. McCarthy, and Mayer Y. Zald, 41–61. New York: Cambridge University Press, 1996.

Taylor, Verta, and Nancy Whittier. "Collective Identity in Social Movement Communities: Lesbian and Feminist Mobilization." In *Frontiers in Social Movement Theory,* edited by Aldon D. Morris and Carol McClurg Mueller, 104–129. New Haven, CT: Yale University Press, 1992.

Thaler, Kai. "Violence Is Sometimes the Answer." *Foreign Policy,* December 5, 2019. https://foreignpolicy.com/2019/12/05/hong-kong-protests-chile-bolivia-egypt-force-police-violence-is-sometimes-the-answer/.

Thompson, Alvin. *Flight to Freedom: African Runaways and Maroons in the Americas.* Kingston, Jamaica: University of West Indies Press, 2006.

Todorov, Tzvetan. *Conquest of America: The Question of the Other.* Translated by Richard Howard. Norman: University of Oklahoma Press, 1999.

Törnberg, Anton. "Combining Transition Studies and Social Movement Theory: Towards a New Research Agenda." *Theory and Society* 47, no. 3 (June 2018): 381–408. https:// doi.org/10.1007/s11186-018-9318-6.

Trouillot, Michel-Rolph. *Silencing the Past: Power and the Production of History.* Boston: Beacon Press, 1995.

Turchin, Peter. "Dynamics of Political Instability in the United States, 1780–2010." *Journal of Peace Research* 49, no. 4 (July 2012): 577–591.

Ture, Kwame. *Black Power: The Politics of Liberation.* New York: Vintage Books, 1967.

Ture, Kwame. "Stokely Carmichael 'We Ain't Going.'" Speech, 1967. https:// americanradioworks.publicradio.org/features/blackspeech/scarmichael.html

Umoja, Akinyele. *We Will Shoot Back: Armed Resistance in the Mississippi Freedom Movement.* New York: New York University Press, 2014.

Unidad Debates Económicos. "The Economic Consequences of the Boycott against Venezuela." Centro Estratégico Latinoamericano de Geopolítica, March 12, 2019. https://www.celag.org/economic-consequences-boycott-against-venezuela/.

United Nations, "Updated Set of principles for the protection and promotion of human rights through action to combat impunity" (E/CN.4/2005/102/Add.1). New York: United Nations, 2005.

United Nations. *Framework for Analysis of Atrocity Crimes: A Tool for Prevention.* New York: United Nations, 2014.

United Nations. "The Rule of Law and Transitional Justice in Conflict and Post-Conflict Societies." Report of the Secretary-General, S/2004/616, 2004. https://digitallibrary. un.org/record/527647?ln=en

United Nations General Assembly. *Charter of Economic Rights and Duties of States: Resolution/Adopted by the General Assembly.* December 17, 1984, A/RES/39/163.

United Nations General Assembly. *Report of the Special Committee on Peacekeeping Operations.* A/72/19. New York: United Nations, 2018.

United Nations General Assembly. "Review of the United Nations Peacebuilding Architecture." Resolution 70/262. New York. April 27, 2016.

United Nations General Assembly. *Universal Declaration of Human Rights.* December 10, 1948.

United Nations Office for the Coordination of Humanitarian Affairs. "Global Humanitarian Overview 2020." December 10, 2019. https://www.unocha.org/sites/ unocha/files/GHO-2020_v9.1.pdf.

United Nations Secretary General. "Statement by the Burundi Configuration of the UN Peacebuilding Commission." May 15, 2015. https://www.un.org/sg/en/content/sg/note-correspondents/2015-05-15/statement-burundi-configuration-un-peacebuilding.

United Nations Security Council. Resolution 1975/2011. New York. March 30, 2011.

United Nations Security Council. Resolution S/RES/2459. New York. March 15, 2019.

U.S. Agency for International Development. "Considerations for Using Data Responsibly at USAID." May 1, 2019. https://www.usaid.gov/responsibledata.

U.S. Department of Health and Human Services, National Commission for the Protection of Human Subjects of Biomedical and Behavioral Research. "The Belmont Report." April 18, 1979. https://www.hhs.gov/ohrp/sites/default/files/the-belmont-report-508 c_FINAL.pdf.

U.S. Department of Health and Human Services, National Institutes of Health. "The Nuremberg Code." https://history.nih.gov/display/history/Nuremberg+Code.

U.S. Department of Health and Human Services, Office for Human Research Protections. "Guidance." February 16, 2016. https://www.hhs.gov/ohrp/regulations-and-policy/ guidance/index.html.

U.S. Embassy in Venezuela. "Secretary Pompeo Remarks to the Press on Venezuela." March 11, 2019. https://uk.usembassy.gov/secretary-pompeo-remarks-to-the-press-on-venezuela/

U.S. White House Office of the Press Secretary. "White House Daily Briefing." Washington, D.C., January 28, 2019. https://www.youtube.com/watch?v=jHyJgnOC-_o

U.S. White House Office of the Press Secretary. "Executive Order: Blocking Property and Suspending Entry of Certain Persons Contributing to the Situation in Venezuela." Washington, D.C., 2015. https://obamawhitehouse.archives.gov/the-press-office/2015/ 03/09/executive-order-blocking-property-and-suspending-entry-certain-persons-c.

Van der Zee, Bibi. "Less than 2% of Humanitarian Funds 'Go Directly to Local NGOs.'" *The Guardian*, October 16, 2016. https://www.theguardian.com/global-developm

ent-professionals-network/2015/oct/16/less-than-2-of-humanitarian-funds-go-direc
tly-to-local-ngos.

Verdeja, Ernesto. "Critical Genocide Studies and Mass Atrocity Prevention." *Genocide Studies and Prevention* 13, no. 3 (2019): 111–127.

Waller, James. *Confronting Evil: Engaging Our Responsibility to Prevent Genocide.* Oxford: Oxford University Press, 2016.

Walters, Kerry. *American Slave Revolts and Conspiracies.* Santa Barbara, CA: ABC-CLIO, 2015.

Warfield, Wallace. "Is It the Right Thing to Do? A Practical Framework for Ethical Decisions." In *A Handbook of International Peacebuilding: Into the Eye of the Storm,* edited by John Paul Lederach and Janice Moomaw Jenner, 213–224. San Francisco, CA: Jossey-Bass, 2002.

Watanabe, Lisa. "UN Mediation in Libya: Peace Still a Distant Prospect." *CSS Analyses in Security Policy,* no. 246 (2019): 1–4.

Wehr, Paul. *Conflict Regulation.* Boulder, CO: Westview Press, 1979.

Weston, Anthony. *Creative Problem-Solving in Ethics.* New York: Oxford University Press, 2007.

Weston, Anthony. *A Practical Companion to Ethics.* 4th ed.. New York: Oxford University Press, 2011.

Weston, Anthony. *A 21st Century Ethical Toolbox.* 3rd ed. New York: Oxford University Press, 2013.

White, Michael, and David Epston. *Narrative Means to Therapeutic Ends.* New York: Norton, 1990.

Whyte, W. H., Jr. "Groupthink." *Fortune* (1952). Groupthink, (Fortune 1952) | Fortune.

Wild, Franz. "ICC Warrant 'Pours Oil on Fire' in Libya, African Union Says." *Bloomberg,* June 29, 2011. http://www.bloomberg.com/news/2011-06- 29/icc-warrant-pours-oil-on-fire-in-libya-african-union-says-1-.htm.

Wilkinson, Steven I. "Riots." *Annual Review of Political Science* 12, no. 1 (2009): 329–343.

Williams, Robert F. *Negroes with Guns.* Eastford, CT: Martino, 1962.

Wines, Michael. "'Looting' Comment from Trump Dates Back to Racial Unrest of the 1960s." *New York Times,* May 29, 2020. https://www.nytimes.com/2020/05/29/us/loot ing-starts-shooting-starts.html.

Yaroshefsky, Ellen. "Military Lawyering at the Edge of the Rule of Law at Guantanamo: Should Lawyers Be Permitted to Violate the Law?" *Hofstra Law Review* 36, no. 2 (2007): 563–600.

Yoo, John. "Bybee Memorandum." U.S. Department of Justice. Memorandum from Jay S. Bybee, Assistant Attorney General, to Alberto R. Gonzales, Counsel to the President, Re: Standards of Conduct for Interrogation under 18 U.S.C. sec.2340–2340A. August 1, 2002. https://www.justice.gov/olc/file/886061/download.

Z, Mickey. "History of the Bonus Expeditionary Force (BEF) or Bonus Army." Zinn Education Project. Accessed November 22, 2020. https://www.zinnedproject.org/materials/bonus-army.

Zelditch, M., Jr., and H. A. Walker. "Legitimacy and the Stability of Authority." *Advances in Group Processes* 1 (1984): 1–26.

Zenko, Micah. "The Big Lie about the Libyan War." *Foreign Policy,* March 22, 2016. https://foreignpolicy.com/2016/03/22/libya-and-the-myth-of-humanitarian-intervention/.

Index

For the benefit of digital users, indexed terms that span two pages (e.g., 52–53) may, on occasion, appear on only one of those pages.